Arithmetic
Clear and Simple

*the text of this book is printed
on 100% recycled paper*

ABOUT THE AUTHOR

Professor Mira holds the degree of Bachelor of Science from Pennsylvania Military College and the degree of Master of Arts from Columbia University. After ten years in industry as an engineer, he began to teach mathematics at Manhattanville College, Purchase, New York, where he has taught for nearly thirty years. In addition, he has been a lecturer in mathematics at St. John's University, Brooklyn, and at the College of the City of New York. He is a Fellow of the American Association for the Advancement of Science, and the author of four other books on mathematics.

EVERYDAY HANDBOOKS

ARITHMETIC
CLEAR and SIMPLE

by

Julio A. Mira

BARNES & NOBLE BOOKS

A DIVISION OF HARPER & ROW, PUBLISHERS

New York, Hagerstown, San Francisco, London

Preface

No one today questions the necessity of developing the intellectual resources of the nation in order to meet the challenge of the modern world. It is a truism to stress the need of a knowledge of arithmetic for there is hardly a national activity which does not depend on a practical understanding of this subject for its successful development. And yet, the average high school graduate knows only a few rules of arithmetic and has only a rudimentary grasp of methods of computation. Most adults—college students, professional and business men and women, even teachers—will admit that they are frequently hampered by an inadequate knowledge of the fundamental concepts of arithmetic. In fact, when difficulty arises in solving a problem, most people suspect the validity of their reasoning process. Yet in most cases, the fault lies in an imperfect understanding of the nature of numbers and a lack of skill in handling them.

The "new" mathematics, that is the new approach to mathematics, as distinguished from the "old" or "traditional" methods of teaching this subject, introduces the most important concepts at an early stage. Many of these concepts and topics are familiar, but the presentation is new and approaches the subject in a basically different manner. One of the most important of these concepts is the idea of a set. Although a thorough knowledge of set theory may be practical only for the specialist, every interested man and women will find the concept well within his grasp and its application to arithmetic interesting and easy to understand.

This book is written for adults. In presenting the new mathematics, it aims to develop an understanding of numbers and their operations so that the reader can improve his computational skill. Every basic principle has been explained and illustrated for the reader. The idea has been to demonstrate the logical character of the principles so clearly that the reader will be able to follow the explanations with ease and be certain that he understands them. The author started preparing this book with the basic proposition that before attempting to apply or teach a subject, one must know that subject thoroughly. All too often even teachers of arithmetic have had little or no opportunity to study this subject from an adult's point of view.

The material presented in this book is intended to meet a wide variety of needs whether experienced from the normal curiosity of the educated man or woman, or arising from the roles of student, parent, teacher, or professional worker. Many examples have been provided in each section. Exercises, with a surprise here and there in the form of a mathematical puzzle, have been carefully constructed with the aim of developing the reader's skill and taste for a subject which did not begin with Pythagoras and did not end with Einstein, but is the oldest and the youngest of all.

J. A. M.

Table of Contents

Number concept. Counting; sets. Cardinal number. Ordinal number. Integers. Old numeral symbols. Roman numerals. Modern system. Bases of number systems.

Fundamental operations. Addition. Laws of addition. Process of addition. Single-digit combinations. Addition of numbers of two or more digits. Rapid addition. Checking addition.

Subtraction. Laws of subtraction. Process of subtraction. Single-digit combinations. Subtraction of numbers of two or more digits. Complementary method of subtraction. Checking subtraction.

Multiplication. Laws of multiplication. Process of multiplication. Single-digit combinations. Multiplication of numbers of two or more digits. Short methods of multiplication. Complementary method of multiplication. Early methods of multiplication. Checking multiplication.

Division. Laws of division. Types of work in division. Single-digit divisors. Divisors of two or more digits. Short methods of division. Early forms of division. Checking division.

List of Symbols

+ Plus, add.

− Minus, subtract.

± Plus or minus.

× Times, multiply.

$\left.\begin{matrix} \div \\ \pm \end{matrix}\right\}$ Divided by, thus, $6 \div 2 = \dfrac{6}{2}$.

$\left.\begin{matrix} \colon \end{matrix}\right\}$ Is to, ratio, thus, $5{:}3 = \dfrac{5}{3}$ is read "5 is to 3."

= Equals, is equal to.

≠ Is not equal to.

$\sqrt[n]{a}$ The nth root of a.

a^{-n} The reciprocal of a^n. Thus $a^{-n} = \dfrac{1}{a^n}$.

( ) Parentheses.

[ ] Brackets.

{ } Braces, the set of.

G. C. D. Greatest common divisor.

L. C. M. Least common multiple.

L. C. D. Lowest common denominator.

% Per cent.

Introduction

Number Concept.* Arithmetic is the science of numbers and begins with the concept of the one and many, that is, of the unit and of plurality. Anthropological studies of primitive peoples indicate that our remote ancestors had a very limited number sense. For example, some of the earliest tribes of Africa, Australia, and Brazil had a number concept limited to one, two, and many. However, no matter how primitive the peoples, there is no language known in which the suggestion of number does not appear. The concept of number seems to be fundamental to the mind of man.

The ability to distinguish the one and the many is not a unique attribute of human beings. According to ornithologists, if a nest contains four eggs, one may safely be taken; but if two are removed, the bird generally deserts. It seems clear that the bird can distinguish two from three. The female "solitary wasp" lays her eggs in individual cells and provides each cell with a number of caterpillars for the young to feed on when they hatch. Some species provide one caterpillar per cell, others five, and still others ten, fifteen, or twenty-four, these numbers being constant according to the species. However, the ability to distinguish quantity should not be confused with counting. Counting involves a specific mental process which is an attribute of human beings exclusively.

Counting; Sets. Fundamentally, the process of counting consists in putting the members of two groups into *one-to-one correspondence*. This principle of one-to-one correspondence may be illustrated as follows. A primitive shepherd could keep track of his flock of sheep by carrying two bags, one full of pebbles and

---

*See Tobias Dantzig, *Number, the Language of Science* (New York: Macmillan, 1939).

the other empty. In the morning, every time a sheep passed through the gate of the pen, the shepherd could transfer a pebble from the bag full of pebbles, say bag *A*, to the empty bag, say bag *B*. When all the sheep were out of the pen there would be one and only one sheep for each pebble in bag *B*, and for each sheep in the flock there would correspond one and only one pebble in bag *B*. That is to say, the pebbles in bag *B* and the sheep in the flock would be in one-to-one correspondence. In general, *two groups of objects are said to be in one-to-one correspondence if the objects of the two groups can be matched in such a way that to every object of the first group there corresponds one and only one object of the second group; and conversely, to every object of the second group there corresponds one and only one object of the first group.*

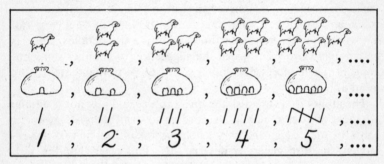

Fig. 1.

Obviously (see Fig. 1), the same result is obtained if instead of pebbles we use sticks, tally marks, or any known set of symbols. Mathematically, *a set is a collection of definite and distinct objects existing either in fact or in thought and determined by the characteristic which they have in common.* For example, the Jones children form a set for they are definite and distinct objects existing in fact and determined by the common characteristic that they have the same parents. The integers 1, 2, 3, 4 also form a set for they are definite and distinct objects existing in thought and determined by the common characteristic that they are integers. Each member of a set is called an *element* of the set. All the months of the year that have 30 days form a set whose elements are April, June, September, and November. The symbol { } is read "the set of." Thus, this set is denoted {April, June, September, November}. We then say that two sets are equal if the elements of one set can be put into one-to-one correspondence with the elements of the other set. For example, the set of all the legs of a horse and the set of all the months of the year that have 30 days are equal, for they can be put into one-to-one correspondence. For each leg of the horse, there corresponds one month that has 30 days and conversely, for each month that has 30 days, there corresponds one leg of the horse.

Cardinal Number. Two sets have the same number of elements if the elements of one set can be put into one-to-one correspondence with the elements of the other set. This number, based on the principle of one-to-one correspondence, is called the *cardinal number* of the set. Thus, in the example above, the cardinal number of both sets is 4 since there are four legs of the horse, each in a one-to-one correspondence with the 4 months of the year that have 30 days. Hence, the cardinal number answers the question "How many?" by matching the elements of a set with those of a model set, but *it does not imply counting.*

Ordinal Number. To count it is necessary to create a number system; that is, the model set must be arranged in an ordered sequence and each successive symbol in the model set must represent a set containing one more element than the elements contained in the set represented by the preceding symbol. Thus, as the primitive shepherd dropped each pebble into bag *B*, it was associated with a sheep. But each pebble as it was dropped into the bag made a set with the other pebbles already in the bag, and the number of that set of pebbles was also the number of the set of sheep which had passed. Thus, we pass from one set to the next by the addition of one more element to the preceding set. This is a concept of order or an *ordinal concept* and gives rise to the *ordinal number.* Hence, an ordinal number is a number that denotes relative position.

It follows that an ordinal number denotes not only the order of the elements of a set but also the cardinal number property of the set. For example, the statement, "Jones ranks number five in his class," denotes not only the position of Jones in the group but also the fact that the group of which Jones is the last member has the cardinal number five.

In order to distinguish between the two concepts, the first three ordinal numbers are frequently written 1st, 2nd, and 3rd and read first, second, and third, respectively. The following ordinal numbers are then formed from the cardinals by adding the suffix *th,* as 4th, 5th, 6th,..., read fourth, fifth, sixth,.... In general, if the words first, second, or third, or the suffix *th* can be applied to a number without changing its meaning, the number is ordinal. Thus, "Jones ranks fifth in his class" does not change the meaning of the above statement. Hence, in this case, the number 5 is ordinal.

Integers.   Both the cardinal and the ordinal aspects of number are necessary for a number system. We definitely must have an answer to "How many?" which is expressed by a cardinal number. But we also need the concept of the ordinal number, for the operations of arithmetic are based on the assumption that we can pass from any number to the succeeding number. Hence, *a number is a collection of units arranged in an ordered sequence.* The units which we collect are whole things. The numbers then are whole numbers or *integers.*

Example.
Figure 2 shows a family sitting at the table. Name sets of:
(a) 4 elements;
(b) 3 elements;
(c) 2 elements;
(d) 1 element;
(e) no elements;
(f) 12 elements.

Fig. 2.

Solution.
(a) The sets containing 4 elements are the sets of persons, plates, knives, forks, spoons, glasses, and chairs.
(b) The set containing 3 elements is the set of males.
(c) The sets containing 2 elements are the set of boys and the set of adults.
(d) The sets containing 1 element are the sets of females, tables, and tablecloths.
(e) An example of a set of no elements, is the set of girls.
(f) A set of 12 elements is the set of silverware.

Old Numeral Symbols.† Just as soon as man acquired property of his own, whether a piece of land, a herd of cattle, or bushels of wheat, he was faced with the necessity of keeping records. The recording of quantity requires some system of numerical symbols or *numerals*. The oldest records showing a systematic use of numerals are those of the ancient Babylonians and Egyptians.

In writing numbers the Babylonians used a vertical wedge to denote 1, the figure ⟨ to represent 10, and the double figure ⟩⊢ to represent 100. The principle of addition was used to denote numbers less than 200. For example, Υ ⊢ ⟨ ⟨ ⟨ Υ Υ represents 132. The principle of multiplication was used for larger numbers. Thus, ⟨Υ ⊢ means 10 times 100 or 1,000. Besides the principles of addition and multiplication, the principle of subtraction was used. For example, some Babylonian tablets show ⟨⟨Υ Υ̅ equal to 19. The symbols ⟨⟨ stand for 20, the symbol Υ⊢ for minus, and Υ for 1, so that ⟨⟨Υ Υ̅ reads 20 minus 1 or 19. The Babylonians used a decimal scale and also a sexagesimal scale. In other words, numbers were grouped in sets of ten (decimal) and also in sets of sixty (sexagesimal). The early Babylonians had no symbol for zero. Sometimes a horizontal line was drawn to denote the absence of units of lower denomination.

EGYPTIAN HIEROGLYPHICS

1	10	100	1,000	10,000	100,000	1,000,000	10,000,000
𝍩	∩	℮	𝍦	𝍧	🐟	𓁨	☉

Fig. 3.

The oldest form of Egyptian numerals is the hieroglyphic, appearing in monuments as far back as 3000 B.C. In hieroglyphic symbols (Fig. 3), the vertical staff ‖ stands for 1, the handle ∩ for 10, the scroll ℮ for 100, the lotus plant 𝍦 for 1,000, a pointing finger 𝍧 for 10,000, the burbot 🐟 for 100,000, the cosmic deity 𓁨 for 1,000,000, and the sun ☉ for 10,000,000. The Egyptians used the principles of addition and multiplication to write

†For a thorough presentation of early number systems, see Florian Cajori, *A History of Mathematical Notations*, Vol. 1 (LaSalle: Open Court, 1928). The symbols have been reproduced by permission of the publisher.

numbers. Not more than four symbols of the same kind were placed in any one group. Thus, four was written ||||, but five was written ||| || or $\overset{|||}{||}$, always putting the larger group before or above the smaller group.

Modern Symbol	Greek Symbol	Name	Modern Symbol	Greek Symbol	Name
1	α	alpha	100	ρ	rho
2	β	beta	200	σ	sigma
3	γ	gamma	300	τ	tau
4	δ	delta	400	υ	upsilon
5	ϵ	epsilon	500	ϕ	phi
6	ς	van	600	χ	chi
7	ζ	zeta	700	ψ	psi
8	η	eta	800	ω	omega
9	θ	theta	900	ϡ	sampi
10	ι	iota	1,000	,α	
20	κ	kappa	2,000	,β	
30	λ	lambda	3,000	,γ	
40	μ	mu	4,000	,δ	
50	ν	nu	5,000	,ϵ	
60	ξ	xi	6,000	,ς	
70	o	omicron	7,000	,ς	
80	π	pi	8,000	,η	
90	ϙ	koppa	9,000	,θ	

Fig. 4.

The Greeks used the twenty-four letters of their alphabet together with the symbols ς van, ϙ koppa, ϡ sampi, and M to represent numbers. As shown in Fig. 4, the first eight letters (alpha to theta) plus van (ς), as number 6, represented the numbers from one to nine, the next eight letters and koppa (ϙ) denoted the tens from ten to ninety, the following eight letters and sampi (ϡ) referred to the hundreds from one hundred to nine hundred. The thousands were represented by the first eight letters and van with the symbol , placed before the letters, as shown in Fig. 4. The symbol M, or Mν, or simply the dot •, was used to denote 10,000. The principle of addition was used. Thus, $\mu\delta$ = 40 + 4 = 44; and ,α ϡ $\iota\eta$ = 1000 + 900 + 10 + 8 = 1,918.

Roman Numerals. The fact that Roman numerals are still in use makes them the most important of all the ancient systems of notation. The symbol I denotes 1; V represents 5; X means 10; L denotes 50; C represents 100; D stands for 500; and M means 1,000. A line placed over a symbol, or symbols, multiplies the value of the symbols by 1,000. Thus $\overline{V}$ denotes 5,000; $\overline{XX}$ means 20,000. Two lines placed over a symbol, or symbols, multiplies the symbols by a million. For example, one million can be written $\overline{\overline{I}}$ or $\overline{M}$, and $\overline{\overline{IV}}$ denotes four millions.

The following rules govern the use of Roman numerals.

1) If a symbol equal in value to, or less than, a given symbol is written to the right of the given symbol, then this symbol is added to the given symbol. Thus, XV means X + V = 10 + 5 = 15, and CC denotes C + C = 100 + 100 = 200.

2) If a symbol less in value than a given symbol is written to the left of the given symbol, then this symbol is subtracted from the given symbol. For example, IV = V − I = 5 − 1 = 4. This principle is used in writing fours and nines only. For example, 99 is not written IC but XCIX, and 4 = IV, 40 = XL, 400 = CD, 9 = IX, 90 = XC, 900 = CM.

3) Not more than three successive equal symbols can be used by themselves or to the right of another greater symbol. Thus, 40 is not written XXXX but XL; 4 is not written IIII but IV; and 9 is not written VIIII but IX. Moreover, not more than one symbol can be written to the left of another greater symbol. For example, 70 is not written XXXC but LXX.

Modern System. The modern system of writing numbers was probably developed by the Hindus of India and introduced to Europe, through Spain and Italy, by the Arabs. Consequently, it is called the Hindu-Arabic numeral system. This method develops naturally from our habit of naming numbers by groups. The fact that finger counting has preceded any other counting technique suggests the reason why numbers are collected in sets of ten. The number of arithmetical symbols needed is then one less than the number of elements in a set, that is, nine. The symbols 1, 2, 3, 4, 5, 6, 7, 8, and 9 are called *digits* from the Latin *digitus* meaning finger. As in any other system, these symbols have an *intrinsic value*, for each represents the number of elements in a fixed set. Thus, 1 represents any one set consisting of one element, 2 denotes any one group containing two objects, 3 represents any one set consisting of triplets, and so forth, passing in order from one set to the next by the addition of one element to the preceding set.

The fundamental principle which makes the Hindu-Arabic system completely different from, and immensely superior to, any other system of notation is the remarkably clever and simple idea of *place value*. This principle is applied by giving each digit a definite value by itself, that is, an intrinsic value, and a *relative value*, determined by its place in the numeral relative to the digit representing the units. Thus, starting with the digit representing units and reading from right to left, the digit written immediately to its left has a value which is ten times greater than its value in the units place and hence, denotes tens; the digit written immediately to the left of the digit in the tens place has a value which is ten times greater than its value in the tens place and thus, denotes hundreds. Figure 5 illustrates the names of the places from their position relative to the units place as used in the United States. (In most of Europe and Latin America a billion is a million millions; in the United States it is a thousand millions.)

	Billions	Hundred Millions	Ten Millions	Millions	Hundred Thousands	Ten Thousands	Thousands	Hundreds	Tens	Units
(a)								5		4
(b)						1	6	1	3	8
(c)					7		4	6		
(d)				3	1	4	2		4	
(e)	4		5			4	2			

Fig. 5.

This relative value of a digit requires a symbol that will denote the absence of a group. For as in (a) of Fig. 5, a number might contain 5 sets of a hundred each and 4 sets of one each but no set of ten each. If this number is written 54, and no symbol is used to indicate the absence of sets of ten each, the digit 5 is made to occupy the place of the sets of ten each and hence, its relative value is changed. It is evident that a symbol to denote the absence of a group is necessary. This symbol, 0, is called *zero*. We then have the ten digits 0, 1, 2, 3, 4, 5, 6, 7, 8, and 9 with which we can write all possible numbers by combining their intrinsic and place values in the numeral by the principle of addition. For example, (a) in Fig. 5 means 5 hundreds + 0 tens + 4 units and is read five hundred and four.

We form our number symbols, or numerals, out of the ten basic digits. These numerals are used to write numbers, whereas words are used to name them. It is important to note that numbers are ideas; the words or figures which represent them are only symbols. Thus XV, read "quindecim" in Latin, and 15, read "fifteen" in English, are entirely different symbols, yet they represent the same number. We can erase the symbols, that is, the numerals XV and 15 or the written words "quindecim" and "fifteen," but we cannot erase the number they represent. For the number they represent is an abstract idea, that is, the concept of fifteen.

In writing numbers, commas are used to separate groups of three digits; then we name each group as if it formed a number of three digits. Thus (b) in Fig. 5, that is 16,138, means 1 ten thousand + 6 thousands + 1 hundred + 3 tens + 8 units and is read sixteen thousand one hundred and thirty-eight; (c) is written 704,600 and is read seven hundred and four thousand six hundred; (d) is written 3,142,040 and is read three million one hundred and forty-two thousand forty; and (e) is written 4,050,042,000 and is read four billion fifty million forty-two thousand.

Bases of Number Systems.   If, as usual, 10 is used as the base, the number 2,345 means

$$2(1{,}000) + 3(100) + 4(10) + 5, \text{ or}$$
$$2(10 \times 10 \times 10) + 3(10 \times 10) + 4(10) + 5.$$

The above notation is awkward. A couple of definitions will allow the development of a more concise and neater way of writing these numbers.

A *factor* is any one of two or more numbers which are multiplied together to form a product. Thus 2 and 3 are factors of 6 because $2 \times 3 = 6$. Similarly, 2, 5, and 7 are factors of 70 because $2 \times 5 \times 7 = 70$. A short way of writing $10 \times 10 \times 10$ is 10^3. The small number 3 written above and to the right of 10 is called an *exponent*. An exponent is a number written to the right of and above another number, called the *base*, to indicate how many times the base is to be taken as a factor. The value assigned to the base with a given exponent is called a *power* of the base. Thus $10^2 = 10 \times 10 = 100$. Here, 10 is the base, 2 is the exponent, and 100 is the power. It is evident that $10^1 = 10$; hence, whenever a number appears without an exponent, that number, by implication, is raised to the first power. It now follows that the number 2,345, written above as $2(10 \times 10 \times 10) + 3(10 \times 10) + 4(10) + 5$, can be written as

$$2(10^3) + 3(10^2) + 4(10) + 5.$$

In general, if a_1 (read *a* subscript one), a_2, and so forth, indicate the digits denoting the number of sets of each group, and if the base 10 is used, any number in a decimal system can be expressed in the form

$$a_n(10^n) + a_{n-1}(10^{n-1}) + \cdots + a_1(10) + a_0$$

where $\cdots$ is read "and so on," n denotes any integer, and the a's represent any digits. Thus, in the number 2,345, $n = 3$, $a_3 = 2$, $a_2 = 3$, $a_1 = 4$, and $a_0 = 5$; hence,

$$2,345 = 2(10^3) + 3(10^2) + 4(10) + 5.$$

This makes sense because in the number 2,345, there are 2 sets of 1,000 each ($10^3 = 10 \times 10 \times 10 = 1,000$); 3 sets of 100 each ($10^2 = 10 \times 10 = 100$); 4 sets of 10 each; and 5 sets of 1 each.

It may be easily concluded, from the discussion in the preceding paragraph, that although we collect numbers in sets of ten, that is, we use ten as a base, any other number can be used as a base. In general, if a number b is used as the base, then any number can be expressed in the form

$$a_n(b^n) + a_{n-1}(b^{n-1}) + \cdots + a_1(b) + a_0.$$

Thus, if we used a septimal system (base 7), we would collect numbers in groups of seven; hence, we would need only the seven digits, 0, 1, 2, 3, 4, 5, and 6. Using the place value principle, the place at the left of the units place would represent sevens, the next place at the left would denote forty-nines ($7^2 = 7 \times 7 = 49$), and so on. Thus in this system seven would be written 10 and a number such as one hundred and sixty-five would be written 324_7, the subscript 7 indicating the base used. Thus,

$$324_7 = 3(7^2) + 2(7) + 4$$
$$= 3(49) + 2(7) + 4$$
$$= 147 + 14 + 4 = 165$$

because the number 165 has 3 sets of (7×7), plus 2 sets of (7), plus 4 sets of (1).

If we use 12 as a base, two extra symbols are needed to denote the necessary twelve digits. Letting Γ denote ten and Σ denote eleven, the twelve digits would be

$$0, 1, 2, 3, 4, 5, 6, 7, 8, 9, \Gamma, \Sigma.$$

Then 10_{12} would mean $1(12) + 0 = 12$. Similarly,

$$4\Gamma_{12} = 4(12) + 10 = 48 + 10 = 58 \text{ and}$$

$$1\Sigma\Gamma_{12} = 1(12^2) + 11(12) + 10 = 144 + 132 + 10 = 286.$$

Example. Express each of the following numbers in the form

$$a_n(10^n) + a_{n-1}(10^{n-1}) + \cdots + a_1(10) + a_0:$$

(a) 345; (b) 15,812; (c) 483,051; (d) 5,078,306.

Solution.
(a) $345 = 3(10^2) + 4(10) + 5.$
(b) $15,812 = 1(10^4) + 5(10^3) + 8(10^2) + 1(10) + 2.$
(c) $483,051 = 4(10^5) + 8(10^4) + 3(10^3) + 0(10^2) + 5(10) + 1.$
(d) $5,078,306 = 5(10^6) + 0(10^5) + 7(10^4) + 8(10^3) + 3(10^2) +$
    $0(10) + 6.$

Example. Express each of the following numbers in the decimal system:
(a) 303_5; (b) 2301_7; (c) $27\Sigma9_{12}$; (d) $1\Sigma0\Gamma_{12}$.

Solution.
(a) $303_5 = 3(5^2) + 0(5) + 3 = 3(25) + 3 = 75 + 3 = 78.$
(b) $2301_7 = 2(7^3) + 3(7^2) + 0(7) + 1 = 2(343) + 3(49) + 0 + 1$
    $= 686 + 147 + 1 = 834.$
(c) $27\Sigma9_{12} = 2(12^3) + 7(12^2) + \Sigma(12) + 9$
    $= 2(1728) + 7(144) + 11(12) + 9$
    $= 3456 + 1008 + 132 + 9 = 4,605.$
(d) $1\Sigma0\Gamma_{12} = 1(12^3) + 11(12^2) + 0(12) + 10$
    $= 1728 + 1584 + 0 + 10 = 3322.$

EXERCISE

Express each of the following numbers in the form

$$a_n(10^n) + a_{n-1}(10^{n-1}) + \cdots + a_1(10) + a_0.$$

1. 649   **2.** 842   **3.** 3,021   **4.** 7,508
5. 12,764   **6.** 301,905   **7.** 628,032   **8.** 4,398,290

Express each of the following numbers in the decimal system.

9. 334_5 **10.** 2342_7 **11.** 110011_2 **12.** 1493_{12}
13. 1142_5 **14.** 3413_7 **15.** 1433_5 **16.** 110100_2
17. $80\Sigma_{12}$ **18.** $\Gamma\Sigma9_{12}$ **19.** 1335_7 **20.** 1010011_2

21. Express the number of weeks in a year to the base 7.
22. Express the number of days in a week to the base 7.
23. Express the number of months in a year to the base 12.
24. Express the number of hours in a day to the base 12.
25. Write fifty. Add zero. Add five. Add one-fifth of eight. The result is called "the greatest thing in the world."
26. An automobile odometer, that is an indicator of distance traveled, is based on 12 instead of 10. Each wheel in the odometer is marked with the figures 0, 1, 2, 3, 4, 5, 6, 7, 8, 9, Γ, Σ. (a) How many miles does the odometer shown at the right register? (b) If this automobile traveled 150 miles more, what would the odometer show?

27. The figures show three odometers based on 12 instead of 10. What will each read after 1 more mile?

(a)   (b)   (c)

28. Using Roman numerals, show that six and six make eleven.
29. Change the following incorrect statements to correct statements by moving a single pin.

(a)   (b)

30. Given 16 matches arranged as follows, take up four matches and move one so as to spell what matches are made of.

□ ◇ ◇ □

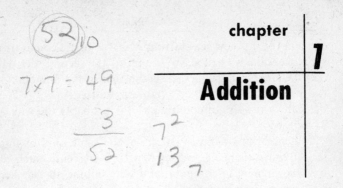

chapter

Addition

1

1.1. Fundamental Operations. A number, as defined on page 4, is a collection of units arranged in an ordered sequence. That is, each number denotes a set of ordered elements. The operations of arithmetic consist in: (a) combining smaller sets to form a larger set and determining the number property of this larger set; or (b) separating a larger set into smaller sets and determining the number property of these smaller sets. *Synthesis* is the combination of separate sets into a whole. *Analysis* is the division of a whole into separate sets. Consequently, the processes of arithmetic are either synthetic or analytic. The operations of addition and multiplication are synthetic processes; and those of subtraction and division are analytic processes. Thus, the four fundamental operations of arithmetic are addition, subtraction, multiplication, and division.

1.2. Addition. Addition is the process of combining smaller sets to form a larger set and determining the number property of this larger set. This process may be visualized as follows. Suppose that we have two sets of the same kind of things, one with 4 elements and the other with 3 elements, and such that they have no element in common. Represent the elements of the first set by four dots {· · · ·} and the elements of the second set by three dots {· · ·}. As shown in Fig. 6, a new, larger set is formed by laying out

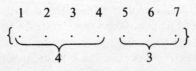

Fig. 6.

the elements of the two sets side by side. Count successively the

13

elements of this new set, starting with the first element of the first set and ending with the last element of the second set. The number of elements in the new set is the *sum* of the numbers denoting the elements of the two original sets, that is, 4 + 3 = 7. This is basically a process of counting and is probably the way addition originated.

In general, if we have two sets of the same kind of things, say set A with *a* elements and set B with *b* elements, such that they have no element in common, and we combine all the elements of both sets into a single, larger set C, we can say that the number of elements *c* in this new set is the sum of the number of elements of the two original sets. Thus, *a* + *b* = *c*. Each of the numbers denoting the elements of the smaller sets, *a* and *b*, is called an *addend*, and the result, *c*, is the *sum*. Moreover, this sum is the number of a new, single set C such that there is one-to-one correspondence between the *a* elements of set A and *a* of the elements of set C, and also between the *b* elements of set B and the remaining elements of set C. Thus, set C contains all the elements which belong to A and to B and to no other set.

Although, as already stated, addition probably started as a process of counting, it is a process of finding the sum of the addends without counting, by using some basic laws and by memorizing the sums of any two digits.

1.3. Laws of Addition. The basic laws of addition are:

1) *Only similar things can be added.* Thus, 3 apples and 4 apples may be added to form a group of 7 apples; but, we cannot add 3 apples to 4 bananas for the result is neither 7 apples nor 7 bananas. We may, of course, consider apples and bananas under the general classification of fruits, and say that the sum of 3 fruits (apples) and 4 fruits (bananas) is 7 fruits.

2) *The sum is a number of things similar to the things added.* Obviously, the sum of 3 chairs and 4 chairs is 7 chairs, not 7 tables or 7 lamps.

3. *The sum is the same regardless of the order in which the numbers are added.* This law follows from the definition of the sum given in the preceding paragraph. As stated above, *a* elements of set C, which is the sum, must be in one-to-one correspondence with the *a* elements of set A, and the *b* elements of set B must be in one-to-one correspondence with the remaining elements of set C. Thus, in the case of 4 + 3 = 7, four elements of the sum 7, are in one-to-one correspondence with the four elements of set 4,

and the three elements of set 3 are in one-to-one correspondence with the remaining elements of the sum (see Fig. 6). It is easy to see that it is just as possible to use the B set first and then the A set, or the 3 set first and then the 4 set, and still obtain a one-to-one correspondence. This law is known as the *commutative law* of addition. If a and b are any two integers, this law may be indicated by writing

$$a + b = b + a.$$

The fact that $2 + 3 = 3 + 2 = 5$ is concretely illustrated in Fig. 7.

COMMUTATIVE LAW OF ADDITION

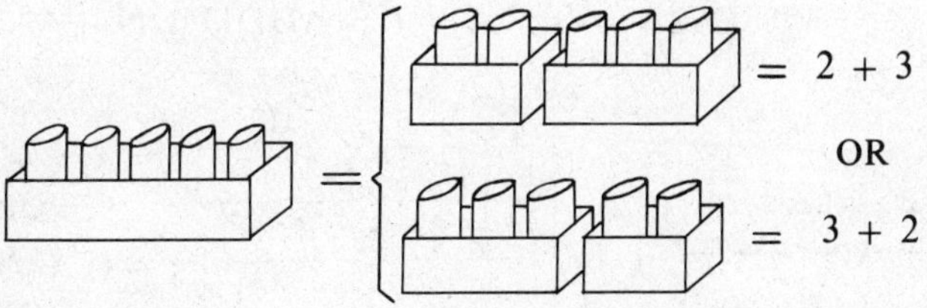

Fig. 7.

Since it is possible to add only two integers at a time, every addition of three or more integers is accomplished by adding the third integer to the sum of the first two, the fourth integer to the sum of the first three, and so on. For example, to find the sum of $1 + 2 + 3$, we can add $1 + 2$ obtaining 3 and then add 3 to 3. Thus,

$$1 + 2 + 3 = 3 + 3 = 6.$$

To avoid any doubt in the grouping of numbers, parentheses ( ), brackets [ ], and braces { } will be used. The above expression is clearer if we write

A $1 + 2 + 3 = (1 + 2) + 3 = 3 + 3 = 6.$

4) *Any method of grouping may be used to obtain the sum of several addends.* If we consider the numbers 1, 2, and 3 in reverse order and form the sum by adding 3, 2, and 1, we can write

B $3 + 2 + 1 = (3 + 2) + 1 = 5 + 1 = 6.$

It then follows that since **A** and **B** are both equal to 6, they are equal to each other, that is, $(1 + 2) + 3 = (3 + 2) + 1$. But, by the commutative law, $3 + 2 = 2 + 3$ and $(3 + 2) + 1 = (2 + 3) + 1 = 1 + (2 + 3)$. Therefore,

$$(1 + 2) + 3 = 1 + (2 + 3).$$

This is the *associative law* of addition. It is an extension of the commutative law and states that the addends may be grouped in any order to obtain the sum. If a, b, and c are any integers, the associative law may be written

$$a + b + c = (a + b) + c = a + (b + c) = b + (a + c).$$

The associative law is illustrated in Fig. 8.

ASSOCIATIVE LAW OF ADDITION

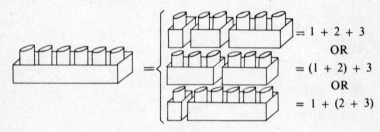

$= 1 + 2 + 3$

OR

$= (1 + 2) + 3$

OR

$= 1 + (2 + 3)$

Fig. 8.

Note that these laws are independent of the notation used to express numbers, or the base of the number system used. For example, $7 + 4 = 4 + 7 = 11$, or $VII + IV = IV + VII = XI$; or using the base 7, $10_7 + 4_7 = 4_7 + 10_7 = 14_7$. Moreover, $II + IV + V = (II + IV) + V = VI + V = II + (IV + V) = II + IX = XI$, or $2 + 4 + 5 = (2 + 4) + 5 = 6 + 5 = 2 + (4 + 5) = 2 + 9 = 11$.

1.4. Process of Addition.   Addition is usually taught in four parts.

1) The memorization of the single-digit combinations.

2) The addition of numbers of two or more digits without carrying.

3) The addition of numbers of two or more digits with carrying.

4) The memorization of decade combinations.

1.5. Single-Digit Combinations. There are a total of one hundred single-digit combinations, see Fig. 9.

ADDITION TABLE

s	0	1	2	3	4	5	6	7	8	9
0	0	1	2	3	4	5	6	7	8	9
1	1	2	3	4	5	6	7	8	9	10
2	2	3	4	5	6	7	8	9	10	11
3	3	4	5	6	7	8	9	10	11	12
4	4	5	6	7	8	9	10	11	12	13
5	5	6	7	8	9	10	11	12	13	14
6	6	7	8	9	10	11	12	13	14	15
7	7	8	9	10	11	12	13	14	15	16
8	8	9	10	11	12	13	14	15	16	17
9	9	10	11	12	13	14	15	16	17	18

Fig. 9

To find the sum of 2 and 3, run your finger down the column headed *s* until you reach 2, then move your finger along the horizontal row corresponding to 2 until you reach the column headed 3. The answer is 5.

If the zero addition is eliminated on the basis that zero added to any number results in the same number, that is, $0 + N = N$, the combinations are reduced to eighty-one. Then it may be argued that since 1 added to any number results in the next highest number, counting by ones takes care of the 1 combinations. There are nine of these sums, leaving seventy-two combinations. Finally, the fact that addition is commutative, that is, $2 + 3 = 3 + 2$, etc., cuts this number in half, leaving thirty-six basic sums. These sums are shown in the table (Fig. 9) within the heavy black lines.

However, a thorough memorization of these thirty-six basic sums is not sufficient. For knowing that 2 + 3 = 5 does not imply that one has the answer to 3 + 2; hence, the one hundred combinations should be learned so there is no hesitation whatever in producing an immediate answer.

Example. Find the sum of: (a) 3 + 3; (b) 7 + 3; (c) 4 + 8; (d) 6 + 7.

Solution. (a) 6. This sum should be seen immediately by simply looking at 3 + 3. *It should not be obtained* by having to say in detail three and three are six. (b) 10. (c) 12. (d) 13.

Example. Find the sum of: (a) 2 + 3 + 4; (b) 3 + 5 + 7; (c) 8 + 4 + 5; (d) 7 + 2 + 6.

Solution.

(a) 9. The sum 2 + 3 + 4 *should not be obtained* by thinking in detail 2 and 3 are 5, and 5 and 4 are 9; but, by immediately seeing the answer 5 to 2 + 3, and just as quickly seeing the answer 9 to 5 + 4, that is, by thinking 5, 9.

(b) 15. Obtained by thinking 8, 15.

(c) 17. Obtained by thinking 12, 17; or from right to left, 9, 17.

(d) 15. Obtained by thinking 9, 15.

Example. Construct, with two pieces of paper, two scales from 1 to 10, as shown below.

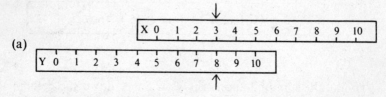

(a)

By placing the scales so that 0 on scale X coincides with 5 on scale Y, we can obtain the sum 5 + 3 = 8 by reading 3 units on the X scale and then reading the corresponding number on the Y scale. Using these scales, show that 3 + 5 = 8.

Solution.

(b)

Example. Using the addition table for a 12 system, shown in Fig. 10, find each of the sums: (a) $9_{12} + 7_{12}$; (b) $5_{12} + 6_{12}$; (c) $\Gamma_{12} + 7_{12}$; (d) $\Sigma_{12} + \Gamma_{12}$.

Solution. (a) 14_{12}; (b) Σ_{12}; (c) 15_{12}; (d) 19_{12}.

Example. Prove each of the following equalities:

(a) $(5 + 4) + 9 = 5 + (4 + 9)$;

(b) $(7_{12} + 6_{12}) + 8_{12} = 7_{12} + (6_{12} + 8_{12})$;

(c) $4_7 + (5_7 + 6_7) = 6_7 + (4_7 + 5_7)$.

Solution.

(a) $9 + 9 = 18$ and $5 + 13 = 18$;

(b) $(7_{12} + 6_{12}) + 8_{12} = 11_{12} + 8_{12} = 19_{12}$;

    $7_{12} + (6_{12} + 8_{12}) = 7_{12} + 12_{12} = 19_{12}$;

(c) $4_7 + (5_7 + 6_7) = 4_7 + 14_7 = 21_7$;

    $6_7 + (4_7 + 5_7) = 6_7 + 12_7 = 21_7$.

ADDITION TABLE
BASE 12

s	0	1	2	3	4	5	6	7	8	9	Γ	Σ
0	0	1	2	3	4	5	6	7	8	9	Γ	Σ
1	1	2	3	4	5	6	7	8	9	Γ	Σ	10
2	2	3	4	5	6	7	8	9	Γ	Σ	10	11
3	3	4	5	6	7	8	9	Γ	Σ	10	11	12
4	4	5	6	7	8	9	Γ	Σ	10	11	12	13
5	5	6	7	8	9	Γ	Σ	10	11	12	13	14
6	6	7	8	9	Γ	Σ	10	11	12	13	14	15
7	7	8	9	Γ	Σ	10	11	12	13	14	15	16
8	8	9	Γ	Σ	10	11	12	13	14	15	16	17
9	9	Γ	Σ	10	11	12	13	14	15	16	17	18
Γ	Γ	Σ	10	11	12	13	14	15	16	17	18	19
Σ	Σ	10	11	12	13	14	15	16	17	18	19	1Γ

Fig. 10.

EXERCISE 1

At sight give the sum of the following numbers.

1. $4 + 1$ **4.** $4 + 4$ **7.** $5 + 4$ **10.** $8 + 8$

2. $2 + 3$ **5.** $3 + 4$ **8.** $3 + 8$ **11.** $5 + 6$

3. $2 + 5$ **6.** $6 + 3$ **9.** $6 + 6$ **12.** $8 + 9$

Find the sum of each of the following.

13. $1 + 2 + 7$ **16.** $7 + 2 + 5$ **19.** $6 + 3 + 4$ **22.** $2 + 4 + 7$
14. $2 + 3 + 6$ **17.** $4 + 3 + 9$ **20.** $4 + 5 + 3$ **23.** $4 + 4 + 7$
15. $3 + 1 + 8$ **18.** $5 + 3 + 8$ **21.** $2 + 5 + 6$ **24.** $6 + 2 + 5$

Using the addition table for a base 12 system shown in Fig. 10, find the sum of the following.

25. $6_{12} + 5_{12}$ **27.** $9_{12} + 7_{12}$ **29.** $\Gamma_{12} + \Gamma_{12}$ **31.** $\Sigma_{12} + 8_{12}$
26. $5_{12} + 7_{12}$ **28.** $7_{12} + 8_{12}$ **30.** $9_{12} + \Gamma_{12}$ **32.** $\Sigma_{12} + \Sigma_{12}$

1.6. Addition of Numbers of Two or More Digits.

a) The addition of numbers of two or more digits in which the combination of two digits is less than ten (that is, without carrying) follows immediately from the single-digit combinations, the principle of place value in our modern system of notation, and the laws of addition. For example, $32 + 24$ states that the sum of

$$32 = 3 \text{ tens} + 2 \text{ units} \quad \text{and}$$
$$24 = \underline{2 \text{ tens} + 4 \text{ units}} \quad \text{is to be found. The result is}$$
$$5 \text{ tens} + 6 \text{ units} \quad \text{by the first law of addition}$$

and the single-digit combinations. Using the principle of place value, this sum is written 56. Thus, the operation is simply written

$$\begin{array}{r} 32 \\ + 24 \\ \hline 56 \end{array}$$

b) The addition of numbers of two or more digits often includes combinations of single digits whose sum is ten or more. Thus, $78 + 84$ states that the sum of

$$78 = 7 \text{ tens} + 8 \text{ units} \quad \text{and}$$
$$84 = \underline{8 \text{ tens} + 4 \text{ units}} \quad \text{is to be found. The result is}$$
$$15 \text{ tens} + 12 \text{ units} \quad \text{by the first law of addition}$$

and the single-digit combinations. Now,

$$12 \text{ units} = 1 \text{ ten} + 2 \text{ units so that}$$
$$15 \text{ tens} + 12 \text{ units} = 15 \text{ tens} + 1 \text{ ten} + 2 \text{ units}$$
$$= 16 \text{ tens} + 2 \text{ units}.$$

Moreover,

$$16 \text{ tens} = 1 \text{ hundred} + 6 \text{ tens hence,}$$
$$16 \text{ tens} + 2 \text{ units} = 1 \text{ hundred} + 6 \text{ tens} + 2 \text{ units}.$$

The sum is found by first adding the units; reducing the resulting sum to tens and units; writing the units of this sum in the units place in the answer (that is, directly under the units column); *carrying* the tens part of the resulting sum to the tens column; adding the tens; and so on. Thus,

EXPLANATION

MENTAL CARRIES: (1)

$$\begin{array}{r} 78 \\ +84 \\ \hline 162 \end{array}$$

$8 + 4 = 12$. Write 2 in the units place and carry 1. Add $7 + 1 = 8$, then $8 + 8 = 16$. Write 6 in the tens place and 1 in the hundreds place.

c) The addition of more than two numbers requires the knowledge of decade combinations. Thus, $8 + 9 + 4 = 17 + 4 = 21$ should be obtained, with practice, by immediately seeing 8, 17, 21. This requires that decade combinations, that is, the combination of a number of more than one digit with a single-digit number, be done mentally and at sight. Decade combinations such as

$$\begin{array}{cccc} 15 & 25 & 65 & 95 \\ \underline{3} & \underline{3} & \underline{3} & \underline{3} \end{array}$$

do not require carrying, for the sum of the units is less than ten. However, combinations such as

$$\begin{array}{ccccc} 5 & 15 & 25 & 65 & 95 \\ \underline{7} & \underline{7} & \underline{7} & \underline{7} & \underline{7} \end{array}$$

require carrying one ten, for the sum of the units is more than ten. Therefore, the tens figure is increased by one.

Thus, in any case, each sum has the units figure obtained from the combination of two digits. If the combination of the two digits in the units place is less than ten, the tens digit of the sum is not increased. If the combination of the two digits is ten or more, the tens digit in the sum is increased by one. For example, the operation $8 + 4 + 5 + 6 = (12 + 5) + 6 = 17 + 6 = 23$ should be done by naming the result of each successive addition, thus, 12, 17, 23.

In general, addition requires the separate summation of the digits denoting units, tens, and so on, and the arrangement of the sum in our modern system of notation. Thus, the addition $1,054 + 432 + 2,363 + 275$ can be performed as follows.

Thousands	Hundreds	Tens	Units
1	0	5	4
	4	3	2
2	3	6	3
	2	7	5

SUM   3   9   21   14

Note that if the sum is written as above, that is, 3 thousands, 9 hundreds, 21 tens, 14 units, the order in which the columns are added is immaterial. It makes no difference whether we add from left to right, or from right to left, or in any order whatever. The arrangement of the sum in our modern system of notation is made by carrying 1 ten from the units column to the tens column to obtain 22 tens, then carrying 2 hundreds from the tens column to the hundreds colunn, and so on, as shown below.

	Thousands	Hundreds	Tens	Units
1st step	3	9	21	14
2nd step	3	9	22	4
3rd step	3	11	2	4
Ans.	4	1	2	4

Thus, the sum is 4,124.

In our modern system of notation, the arrangement of the sum is done, immediately following the summation of each column, by carrying. Hence, we add from right to left, beginning at the units column.

EXPLANATION

MENTAL CARRIES: (1) (2) (1)

```
       1,  0  5  4
           4  3  2
       2,  3  6  3
           2  7  5
      _____
SUM    4,  1  2  4
```

Units column, 6, 9, 14. Write 4 in the units column of the sum and carry 1. Tens column, 6, 9, 15, 22. Write 2 in the tens column and carry 2. Hundreds column, 6, 9, 11. Write 1 in the hundreds column and carry 1. Thousands column, 2, 4.

EXERCISE 2

Perform each of the following additions at sight.

1. 47	**2.** 53	**3.** 63	**4.** 62	**5.** 25	**6.** 34
32	46	24	37	62	49

7. 59	**8.** 28	**9.** 47	**10.** 78	**11.** 69	**12.** 85
39	58	69	77	76	59

13. $8 + 7 + 9 + 6$ **14.** $3 + 7 + 6 + 4$ **15.** $2 + 8 + 5 + 5$

16. $9 + 1 + 3 + 2$ **17.** $5 + 7 + 7 + 4$ **18.** $8 + 5 + 9 + 3$

Perform each of the following additions.

19. 86	**20.** 46	**21.** 64	**22.** 739	**23.** 739	**24.** 234
74	85	73	266	672	743
92	38	39	845	981	754
68	75	88	934	596	369
320	*244*	*264*	*2784*	*2988*	*2100*

Find the sum of each of the following to the indicated base.

25. 96_{12}	**26.** $9\Sigma_{12}$	**27.** $\Gamma\Gamma_{12}$	**28.** 32_7	**29.** 13_7	**30.** 66_7
$2\Gamma_{12}$	$\Gamma 4_{12}$	$\Sigma\Sigma_{12}$	51_7	64_7	66_7
4					

Add each of the following vertically and horizontally, and check the sums by finding the totals of each row and of each column.

31. $46 + 29 + 13 + 96 =$ *184*   **32.** $43 + 73 + 79 + 26 =$ *221*
      $43 + 71 + 65 + 84 =$ *263* $23 + 70 + 42 + 24 =$ *159*
      $28 + 49 + 17 + 64 =$ *158* $61 + 30 + 69 + 38 =$ *188*
      $98 + 67 + 15 + 65 =$ *245* $18 + 32 + 96 + 35 =$ *151*
    215 +216 +110 +309 = 850   *135 +205 +286 + 123 = 749*

33. $62 + 87 + 98 + 24 =$ *271*   **34.** $44 + 24 + 79 + 96 =$ *243*
      $12 + 56 + 97 + 43 =$ *208* $41 + 76 + 76 + 37 =$ *230*
      $79 + 28 + 23 + 99 =$ *229* $94 + 52 + 81 + 99 =$ *326*
      $26 + 74 + 13 + 38 =$ *151* $63 + 48 + 77 + 65 =$ *253*
    179 +245 +331 +204 = 859   *242 200 +313 +297 = 1052*

1.7. Rapid Addition. The application of the associative and commutative laws helps in speeding up the operation of addition. For example, we can perform the addition shown at the left more

     8   quickly by immediately seeing the combinations of 8 and
     6   2 and 6 and 4. The associative law, that is, the fact that
     2   numbers may be grouped in any order to be added, is then
     4   applied. Thus, we think 10 (8 and 2), 20 (10, and 6 and 4),
     5   25 (20 and 5). The expressions in parentheses are written
   25   as an explanation. Actually the addition should be thought
of as 10, 20, 25. Hence, the combinations that give 10, such as
(1, 9), (2, 8), (3, 7), (4, 6), and (5, 5), and those that give 15, such
as (6, 9) and (8, 7), should be immediately recognized.

    When a short two-digit column is to be added, the work can
be speeded up by the following device. The addition shown at

46   the left can be done by thinking 46 (and 60), 106 (and 4),
64   110 (and 20), 130 (and 3), 133 (and 20), 153 (less 2), 151.
23   The expressions in parentheses are again written as an
18   explanation.   Thus, the addition should be actually
151 thought: 46, 106, 110, 130, 133, 153, 151.

1.8. Checking Addition.   The most usual check of addition is the
application of the commutative and associative laws, that is,
adding the numbers in a different order.   Thus, the preceding
two-digit-column addition can be checked by thinking 18 (and
20), 38 (and 3), 41 (and 60), 101 (and 4), 105 (and 40), 145 (and 6),
151.

The so-called Civil Service method of checking is carried out by
adding the units column, then the tens column, etc., and finally
adding each of the results.   The following example illustrates the
method.

	CHECK	
3,425	3,425	
5,917	5,917	
231	231	
4,482	4,482	
14,055	15	units
	14	tens
	1 9	hundreds
	12	thousands
	14,055.	*Ans.*

The method of checking by "casting out 9's" is based on the
fact that the remainder obtained when a number is divided by 9 is
equal to the sum of its digits.*   For example, if the number 321
is divided by 9 the remainder is 6.   The sum of the digits, $3 + 2 +
1$, is also 6.   This number, 6, is called the *check* number of 321.
Moreover, if each addend is divided by a given number, the sum
of the check numbers obtained is equal to the check number obtained by dividing the sum by the same given number.†   Thus, the
above addition can be checked by casting out 9's as follows.

3,425	$3 + 4 + 2 + 5 = 14$ and $1 + 4 =$	5
5,917	$5 + 9 + 1 + 7 = 22$ and $2 + 2 =$	4
231	$2 + 3 + 1 = \ldots\ldots\ldots\ldots\ldots =$	6
4,482	$4 + 4 + 8 + 2 = 18$ and $1 + 8 =$	9
14,055		24

*Proof of this statement is given on page 159.
†Proof of this statement is given on page 159.

The sum of the check numbers of the addends is 24 and $2 + 4 = 6$. The sum, 14,055, has a check number equal to $1 + 4 + 0 + 5 + 5 = 15$ and $1 + 5 = 6$. Since the check number of the sum is equal to the sum of the check numbers of the addends, the addition is correct.

This system of checking has a disadvantage. An error resulting from reversing the order of two figures, that is, of transposition, cannot be discovered. For example, the check number of 321 is $3 + 2 + 1 = 6$, which is the same as the check number of 312. Since this error is made more frequently than supposed, casting out 9's is not a conclusive test.

To detect an error of transposition, the operation can be checked by "casting out 11's." To cast out 11's, start with the figure in the units place and add to it the alternate figures (the figure in the hundreds place, then the figure in the ten thousands place, etc.); then start with the figure in the tens place and add to it the alternate figures (the figure in the thousands place, then the figure in the hundred thousands place, etc.). Finally, subtract the second sum from the first sum.‡ For example, the check number of 67,435 is obtained as follows.

$$5 + 4 + 6 = 15$$
$$3 + 7 = \underline{10}$$
$$\overline{5.} \quad \textit{Check Number.}$$

If the second sum is larger than the first sum, increase the latter by 11 or any multiple thereof. For example, the check number of 14,055 can be obtained as follows.

$$5 + 0 + 1 = 6$$
$$5 + 4 = 9$$

Increase 6 by adding 11, making it 17, then subtract 9. The check number is then $(6 + 11) - 9 = 17 - 9 = 8$.

The preceding addition can now be checked by casting out 11's.

3,425	$(5 + 4) - (2 + 3) = 9 - 5 = 4$
5,917	$(7 + 9) - (1 + 5) = 16 - 6 = 10$
231	$(1 + 2) - (3) = 0$
4,482	$(2 + 4) - (8 + 4) = 6 - 12 \text{ or } (6 + 11) - 12$
	$= 17 - 12 = 5$
14,055	The sum of the check numbers is $4 + 10 + 0 + 5 = 19 \text{ or } 9 - 1 = 8$.

‡ Proof of this statement is given on page 160.

The check number of the sum, 14,055, is $(5 + 0 + 1) - (5 + 4) = 6 - 9$ or $(6 + 11) - 9 = 17 - 9 = 8$. Since the check number of the sum is equal to the sum of the check numbers of the addends, the addition is correct.

EXERCISE 3

(a) Using short methods, find the sum of each of the following. Check the result by casting out (b) 9's; (c) 11's.

1. 6	2. 7	3. 9	4. 23	5. 32	6. 59
3	1	9	42	13	34
7	8	5	34	41	15
4	9	6	51	52	28
2	5	1			
22	*30*	*30*	*150*	*138*	*136*

7. 45	8. 4,732	9. 6,496	10. 2,874	11. 2,541	12. 3,258
56	3,834	3,579	5,781	1,796	2,719
14	1,492	5,853	1,286	4,369	1,417
55	5,837	4,286	4,329	3,514	3,626
170	*15895*	*20214*	*14270*	*12220*	*11R*

13. Find the cost of making seven pieces of chain with four links each into a continuous chain of twenty-eight links, if it costs a cent to open a link and a cent to weld it.

14. A man has a mouse, a cat, and a dog to carry across a river, one by one. In what manner can this be done so that the dog will not be left with the cat nor the cat with the mouse?

15. A snail crawling up a pole 11 feet high climbs up 5 feet each day and slips back 4 feet each night. How long will it take the snail to reach the top?

16. Which is correct to say, 9 and 5 are 13, or 9 and 5 is 13?

17. If there are more than 366 persons in a room, how can you be certain that at least two of them have the same birthday?

Subtraction

2.1. Subtraction. Addition was defined, in Section 1.2, as the process of combining two smaller sets into a larger set and determining the number property of the larger set. Thus, the process of addition is performed when the number property of each of the two smaller sets is given and the number property of the larger set must be determined. The process of subtraction arises when the number property of one of the two smaller sets and the number property of the larger set are given, and the number property of the other smaller set must be determined. Thus, *subtraction* is the process of finding the number property of a set which when added to the smaller of two given sets will give the larger. The two numbers denoting the given sets are called the *minuend* (from *minuendus*, to be lessened) and the *subtrahend* (quantity to be subtracted or taken away). The number denoting the resulting set is called the *difference*. In other words, the difference between two numbers is a number which when added to the smaller number (subtrahend) will equal the greater number (minuend). Evidently, since the difference has to be added to the subtrahend to obtain the minuend, the subtrahend must be either equal to or smaller than the minuend. For example, suppose that we have two sets of the same kind of things, one with 7 elements and the other with 4 elements, and such that they have no element in common. Represent the elements of the first set by seven dots {·······} and the elements of the second set by four dots {····}. We now place the smaller set below the larger set

$$\{ \cdot \ \cdot \ \cdot \ \cdot \ \cdot \ \cdot \ \cdot \}$$
$$\updownarrow \ \updownarrow \ \updownarrow \ \updownarrow$$
$$\{ \cdot \ \cdot \ \cdot \ \cdot \}$$

so that each of the elements of the smaller set corresponds to a particular element of the larger set and then we count, successively, the remaining elements of the larger set. The number of elements in this new set {···} is the difference between the numbers denoting the elements of the two original sets, that is, 7 − 4 = 3. Here, 7 is the minuend, 4 is the subtrahend, and 3 is the difference.

Subtraction is then the inverse operation of addition. As explained in Section 1.1, addition is a synthetic process because it combines smaller sets to form a larger set, whereas subtraction is an analytic process because it divides a larger set into its component smaller sets.

Subtraction can be considered in three ways:

1) As a process that determines the number which must be added to the smaller of two given numbers to obtain the larger. It answers the question, what must be added to a to obtain c?

2) As a process that determines the quantity which must be taken away from the larger of two given quantities to obtain the smaller. It answers the question, what is the result of taking a away from c?

3) As a process that compares two given quantities. It answers the question, how much more is c than a?

The first of these three cases makes use of the addition combinations; hence, it requires no new tables and uses processes which are already familiar. However, to understand the process of subtraction it is necessary to realize that subtraction is a taking-away process. That is, a certain quantity must be taken away from the minuend (or quantity to be lessened) in order to obtain the difference.

2.2. Laws of Subtraction. Since subtraction has been defined as the inverse process of addition, the laws of addition must also hold for subtraction.

1) *Only similar things can be subtracted.* The difference between 3 units and 9 units is 6 units; but, 3 units cannot be subtracted from 9 tens directly. The tens must be expressed in terms of units before the operation can be performed. Thus, 9 tens is equal to 8 tens and 10 units, and 3 units subtracted from 10 units leaves 7 units. Hence, the result of taking 3 units away from 9 tens is 8 tens and 7 units or 87 units. Note that the 3 units were subtracted from 10 units (a similar quantity) and not from 9 tens.

2) *The difference is a number of things similar to the things of the minuend and the subtrahend.* Thus, 3 dollars taken away from 9 dollars leaves 6 dollars, and not pounds or francs.

Subtraction is not commutative for $6 - 2$ is not equal to $2 - 6$. In fact, the latter operation is impossible without the use of both positive and negative numbers. Moreover, subtraction is not associative for $(9 - 5) - 3 = 1$ is not equal to $9 - (5 - 3) = 7$. The next two properties of subtraction are similar to the commutative and associative laws and for this reason they are sometimes erroneously confused with these laws.

3) *Either of the two smaller sets can be taken as the subtrahend and then the other will be the difference.* Thus, if a, b, and c represent any integers and if $a + b = c$, then $c - a = b$ and $c - b = a$. For example, if $6 + 3 = 9$, then $9 - 6 = 3$ and $9 - 3 = 6$.

4) *The result of adding two given numbers and then subtracting this sum from a third larger number is the same as the result of subtracting one of the given numbers from the third larger number and then subtracting, from this difference, the other given number.* Thus, if a, b, and c denote any integers and $b + c$ is less than a, then $a - (b + c) = (a - b) - c$. For example, given 9, 3, and 2, then $9 - (3 + 2) = (9 - 3) - 2 = 4$.

5) *If the same quantity is added to the minuend and to the subtrahend, or is subtracted from the minuend and from the subtrahend, the difference remains unchanged.* Thus, if a, b, c, and d are any integers and if $a - b = c$, then $(a + d) - (b + d) = c$ and $(a - d) - (b - d) = c$ (if d is less than a and less than b). For example, $9 - 5 = 4$, so $(9 + 3) - (5 + 3) = 12 - 8 = 4$ and $(9 - 3) - (5 - 3) = 6 - 2 = 4$.

As in the case of addition, these laws are independent of the notation used to express numbers or the base of the number system used. In other words, since $8 + 3 = 11$, then $11 - 8 = 3$ and $11 - 3 = 8$; and $12 - (4 + 3) = (12 - 4) - 3 = 5$ regardless of the base used or how these numbers are represented.

2.3. Process of Subtraction. Subtraction can be considered:

a) As a process that determines the number which must be added to the smaller of two given numbers to obtain the larger. This is usually called the *additive* process. Thus, to subtract 3 from 5 one must find the number which added to 3 gives 5. Since 2 added to 3 gives 5, then 2 is the result of subtracting 3 from 5. As pointed out in Section 2.1, this method makes use of the addition combinations; hence, it requires no new tables and uses processes which are already familiar.

b) As a process that determines the quantity which must be

taken away from the larger of two given quantities to obtain the smaller. This process is called the *take-away* method. Thus, to subtract 3 from 5 one must find the quantity that remains if 3 is taken away from 5. If, from a set containing 5 elements {·····}, 3 elements are taken away {·/·//}, the remaining set {··} will contain 2 elements. Hence 2 is the result of taking 3 away from 5.

2.4. Single-Digit Combinations. The subtraction of one digit from another equal or larger digit offers no difficulty, for the subtraction combinations are obtained from the already familiar addition combinations.

The single-digit combinations are given on the Subtraction

SUBTRACTION TABLE
Subtrahend

d	1	2	3	4	5	6	7	8	9
0	1	2	3	4	5	6	7	8	9
1	2	3	4	5	6	7	8	9	10
2	3	4	5	6	7	8	9	10	11
3	4	5	6	7	8	9	10	11	12
4	5	6	7	8	9	10	11	12	13
5	6	7	8	9	10	11	12	13	14
6	7	8	9	10	11	12	13	14	15
7	8	9	10	11	12	13	14	15	16
8	9	10	11	12	13	14	15	16	17
9	10	11	12	13	14	15	16	17	18

Fig. 11.

Table to the left of the heavy black lines (Fig. 11). The first row gives the subtrahend and the first column at the left, headed d, gives the difference. Hence, the minuend is found in the body of the table. To obtain the difference between 3 and 5, find the column headed 3; run down this column until the minuend 5 is found; move the finger to the left along this horizontal row until the column headed d is reached. The difference is 2. As in the case of addition, these combinations should be learned so that there is no hesitation whatever in producing an immediate answer.

The difference between two numbers can be found by using either of the two methods discussed in Section 2.3. Thus, we can find the difference resulting from the subtraction of 3 from 5 by:

a) The additive process. Thinking 2 added to 3 gives 5; hence, the difference is 2.

b) The take-away process. Thinking 3 taken from 5 leaves 2; hence, the difference is 2.

EXERCISE 4

At sight, give the difference between each of the following.

1. 4 − 1	**7.** 7 − 4	**13.** 10 − 6	**19.** 11 − 6
2. 3 − 2	**8.** 9 − 6	**14.** 11 − 7	**20.** 17 − 8
3. 3 − 3	**9.** 10 − 5	**15.** 12 − 8	**21.** 16 − 9
4. 4 − 2	**10.** 10 − 2	**16.** 18 − 9	**22.** 13 − 8
5. 7 − 2	**11.** 11 − 9	**17.** 13 − 4	**23.** 14 − 5
6. 6 − 4	**12.** 12 − 3	**18.** 13 − 9	**24.** 15 − 9

Using the addition table for a 12 system shown in Fig. 10, page 19, find the difference between each of the following.

25. $\Gamma_{12} - 4_{12}$	**27.** $10_{12} - \Gamma_{12}$	**29.** $17_{12} - 9_{12}$	**31.** $14_{12} - 9_{12}$
26. $13_{12} - 8_{12}$	**28.** $\Sigma_{12} - 4_{12}$	**30.** $14_{12} - 7_{12}$	**32.** $1\Gamma_{12} - \Sigma_{12}$

2.5. Subtraction of Numbers of Two or More Digits. The subtraction of numbers of two or more digits in which the number in the subtrahend is less than the corresponding number in the minuend is simply a repetition of the single-digit combinations. Either the additive or the take-away method may be used. For example, the subtraction of 623 from 857 can be written as follows.

$$857 = 8 \text{ hundreds} + 5 \text{ tens} + 7 \text{ units}$$
$$\underline{623 = 6 \text{ hundreds} + 2 \text{ tens} + 3 \text{ units}}$$
$$234 = 2 \text{ hundreds} + 3 \text{ tens} + 4 \text{ units}$$

The answer is obtained by using the appropriate laws of subtraction and the single-digit combinations.

a) If the additive method is used, think 3 and 4 give 7 (write 4 in the units column); 2 and 3 give 5 (write 3 in the tens column); 6 and 2 give 8 (write 2 in the hundreds column). As in the case of addition, the answer should be obtained by immediately seeing 4, 3, 2. The process can be visualized by reading each column upward.

SOLUTION	EXPLANATION		
857	$\frac{8}{6}$	$\frac{5}{2}$	$\frac{7}{3}$
623	and	and	and
234	2	3	4

b) If the take-away method is used, think 3 from 7 leaves 4 (write 4 in the units column); 2 from 5 leaves 3 (write 3 in the tens column); 6 from 8 leaves 2 (write 2 in the hundreds column). If this method is used, the single-digit combinations given in the Subtraction Table (Fig. 11) should be memorized and the answer should be immediately seen as 4, 3, 2; that is, 234. This process can be visualized by reading each column downward.

SOLUTION	EXPLANATION		
857	6	2	3
623	from	from	from
234	$\frac{8}{2}$	$\frac{5}{3}$	$\frac{7}{4}$

If the number in the subtrahend is greater than the corresponding number in the minuend, the subtraction is impossible unless the number in the minuend is made greater than the corresponding number in the subtrahend without changing the difference. This may be accomplished in two ways: (a) the *balancing-additions* method and (b) the *borrowing* method. Each of these two methods can be combined with the additive and the take-away methods, resulting in four methods of subtraction: (1) the additive balancing-additions method, sometimes called the Austrian method, (2) the take-away balancing-additions method, (3) the take-away borrowing method, and (4) the additive borrowing method.

To illustrate these methods consider the subtraction of 47 from 82.

1) The additive balancing-additions method.

SOLUTION	EQUIVALENT OPERATION	EXPLANATION

$$\begin{array}{cc} 82 & \quad 8\text{ tens} + 12\text{ units} \\ \underline{47} & \quad \underline{5\text{ tens}} + \underline{7\text{ units}} \\ 35 & \quad 3\text{ tens} + 5\text{ units} \end{array}$$

$$\begin{array}{cccc} & \underline{8} & (2+10) = & \underline{12} \\ & \overline{5} & = (4+1) & \overline{7} \\ & \text{and} & & \text{and} \\ & 3 & & 5 \end{array}$$

Beginning at the right, we reason: 7 units cannot be taken away from 2 units; hence, we add 1 ten (10 units) to the 2 units, obtaining 12 units, and making the minuend 8 tens and 12 units. But if we add 1 ten to the minuend, we must add the same quantity, 1 ten, to the subtrahend so that the difference will remain unchanged (law 5, page 29). Thus, 4 tens and 1 ten gives 5 tens. As shown in the equivalent operation, the answer is then immediately seen as 35. Once the reasoning is understood, the answer should be obtained (without writing the equivalent operation) by thinking: 5 and 7 give 12, carry 1, and 4 gives 5, and 3 gives 8. The difference is 35.

2) The take-away balancing-additions method.

SOLUTION	EQUIVALENT OPERATION	EXPLANATION

$$\begin{array}{cc} 82 & \quad 8\text{ tens} + 12\text{ units} \\ \underline{47} & \quad \underline{5\text{ tens}} + \underline{7\text{ units}} \\ 35 & \quad 3\text{ tens} + 5\text{ units} \end{array}$$

$$\begin{array}{cccc} & 5 & = (4+1) & 7 \\ & \text{from} & & \text{from} \\ & \underline{8} & (2+10) = & \underline{12} \\ & \overline{3} & & \overline{5} \end{array}$$

Again, since 7 units cannot be taken away from 2 units, we add 1 ten (10 units) to the 2 units, obtaining 12 units, and making the minuend 8 tens and 12 units. But, if we add 1 ten to the minuend, we must add the same quantity, 1 ten, to the subtrahend, so that the difference will remain unchanged (law 5, page 29). Thus 4 tens and 1 ten gives 5 tens. The equivalent operation is shown above and the answer is immediately seen to be 35. The answer should be obtained by thinking: 7 from 12 leaves 5; carry 1, and 4 gives 5; then 5 from 8 leaves 3; hence, the difference is 35.

3) The take-away borrowing method.

SOLUTION	EQUIVALENT OPERATION	EXPLANATION

$$\begin{array}{cc} 82 & \quad 7\text{ tens} + 12\text{ units} \\ \underline{47} & \quad \underline{4\text{ tens}} + \underline{7\text{ units}} \\ 35 & \quad 3\text{ tens} + 5\text{ units} \end{array}$$

$$\begin{array}{cccc} & 4 & & 7 \\ & \text{from} & & \text{from} \\ (8-1) = & \underline{7} & (2+10) = & \underline{12} \\ & 3 & & 5 \end{array}$$

Again, 7 units cannot be taken away from 2 units, so we "borrow" 1 ten (10 units) from the 8 tens, leaving 7 tens. Then we add the 10 units to the 2 units, obtaining 12 units. The minuend is now written as 7 tens and 12 units. As shown in the equivalent operation, the answer is immediately seen to be 35. This answer should be obtained by thinking: 82 is equivalent to 7 tens and 12 units; 7 from 12 leaves 5, and 4 from 7 leaves 3. The difference is 35.

4) The additive borrowing method.

SOLUTION	EQUIVALENT OPERATION		EXPLANATION	
82	7 tens +	12 units	$(8-1) = \underline{7}$	$(2+10) = \underline{12}$
47	4 tens +	7 units	4	7
⎯	⎯⎯⎯	⎯⎯⎯	and	and
35	3 tens +	5 units	3	5

Since 7 units cannot be taken away from 2 units, we will again "borrow" 1 ten (10 units) from the 8 tens, leaving 7 tens; and add the 10 units to the 2 units, obtaining 12 units. The minuend is now thought of as 7 tens and 12 units. The equivalent operation is shown above, and as before, the answer is immediately seen to be 35. This answer is obtained directly by thinking: 82 is equivalent to 7 tens and 12 units; 5 and 7 gives 12, and 3 and 4 gives 7; hence, the difference is 35.

Example. Subtract 478 from 732 using each of the four methods.
Solution.
Additive balancing-additions method.

732	8 and 4 = 12. Carry 1, 7 + 1 = 8.
478	8 and 5 = 13. Carry 1, 4 + 1 = 5.
254	5 and 2 = 7.

Take-away balancing-additions method.

732	8 from 12 = 4. Carry 1, 7 + 1 = 8.
478	8 from 13 = 5. Carry 1, 4 + 1 = 5.
254	5 from 7 = 2.

Take-away borrowing method.

732	8 from 12 = 4. Borrow 1 from 3 leaving 2.
478	7 from 12 = 5. Borrow 1 from 7 leaving 6.
254	4 from 6 = 2.

Additive borrowing method.

732	8 and 4 = 12. Borrow 1 from 3 leaving 2.
478	7 and 5 = 12. Borrow 1 from 7 leaving 6.
254	4 and 2 = 6.

EXERCISE 5

Find the difference between each of the following.

1. 72	**2.** 63	**3.** 93	**4.** 74	**5.** 76
48	38	59	39	27
24	*25*	*34*	*35*	*49*
6. 90	**7.** 63	**8.** 852	**9.** 943	**10.** 721
13	25	616	348	644
77	*38*	*236*	*595*	*77*
11. 630	**12.** 901	**13.** 800	**14.** 707	**15.** 6,538
262	188	274	599	2,463
368	*713*	*726*	*108*	*4075*
16. 4,500	**17.** 9,827	**18.** 3,357	**19.** 8,020	**20.** 9,372
1,703	2,854	2,649	3,928	3,986
2797	*6973*	*708*	*4092*	*5386*

21. John owes Bill $914. If John pays Bill $205 of this debt, how much will he still owe Bill? *709*

22. Jim deposits $3,572 in his bank. He withdraws $109; later, he withdraws $657; still later, $798; and finally, $1,108. What is his balance? *900*

23. Mr. Farrell sold his house for $15,680 making a profit of $2,782. How much did Mr. Farrell pay for this house? *12,898*

24. Mr. Moore bought a house for $12,500 and an automobile for $2,980. He sold the house for $15,000 and the automobile for $1,200. How much did Mr. Moore gain or lose in these transactions? *720*

25. A man with $100 wanted $125. He pawned the $100 for $75 and then sold the pawn ticket for $50. He then had $125. Who lost in the transaction?

2.6. Complementary Method of Subtraction. The *arithmetic complement*, or simply the *complement*, of a number is the difference between the number and the next largest power of 10. Thus, the complement of 98 is 2, because 2 is the difference between 100 and 98. The complement of a number can easily be found by subtracting the first digit on the right that is other than zero, from 10, and all other digits from 9. Thus to obtain the complement of 240, subtract 4, which is the first digit on the right that is other than zero, from 10, obtaining 6; then 2 from 9 leaves 7. The result 760 is the complement of 240. Similarly, the complement of 13,420 can be obtained, starting at the left, by thinking $9 - 1 = 8$, $9 - 3 = 6$, $9 - 4 = 5$, and $10 - 2 = 8$. The result 86,580 is the complement of 13,420.

Recalling the principle—if the same quantity is added to the minuend and to the subtrahend the difference remains unchanged —it is easy to see that adding the complement of the subtrahend to both the minuend and the subtrahend leaves the difference unchanged. For all practical purposes then, the complementary method of subtraction reduces subtraction to addition. This fact is illustrated in the following examples.

Example. Subtract 7 from 34.
Solution. The complement of 7 is 3. If 3 is added to 7 and to 34, the difference remains unchanged.

$$34 - 7 = (34 + 3) - (7 + 3) = 37 - 10 = 27.$$

Note that we added 3, the complement of 7, to 34 and then subtracted 10, since it is the next power of $10(10^1)$ that is larger than 7.

Example. Subtract 83 from 546.
Solution. The complement of 83 is 17 since $9 - 8 = 1$ and $10 - 3 = 7$. Hence,

$$546 - 83 = (546 + 17) - (83 + 17) = 563 - 100 = 463.$$

Note that we added 17, the complement of 83, to 546 and then subtracted 100, the next power of 10 (10^2) that is larger than 83.

To obtain the difference by the complementary method, add the complement of the subtrahend to the minuend and then subtract, from this sum, the next power of 10 that is larger than the subtrahend.

Thus, to subtract 7 from 34, we should think 34, 37, 27. The complement should not be written; its successive digits should be carried on mentally. Similarly, to subtract 83 from 546, we

546
 83
463

should think 6 and 7 $(10 - 3)$ is 13. Write 3; carry 1, and 4 gives 5, and 1 $(9 - 8)$ gives 6. Then 1 hundred (the power of 10 larger than 83) subtracted from 5 hundred leaves 4 hundred. Hence, the answer is 463.

Example. From the sum of 643, 274, and 946, subtract 538.
Solution. Arrange the numbers as shown below. The complement of 538 is not written but its successive

MENTAL CARRIES:	(2)	(1)	
	6	4	3
	2	7	4
	9	4	6
	5	3	8
	13	2	5

digits are carried on mentally, as necessary. Since the first digit of the complement of 538 is $10 - 8 = 2$, to add the first column we should think 3, 7, 13, 15. Write 5 and carry 1. To add the second column think 1, 5, 12, 16, 22 (the second digit of the complement of 538 is $9 - 3 = 6$). Write 2 and carry 2. To add the third column think 2, 8, 10, 19, 23 (the third digit of the complement of 538 is $9 - 5 = 4$). Write 3 and subtract 1 from 2, obtaining the answer 1,325.

EXERCISE 6

Find the complement of each of the following.

1. 17   **2.** 39   **3.** 687   **4.** 713

5. 530   **6.** 3,408   **7.** 4,073   **8.** 6,500

Using the complementary method, perform the indicated operations.

9. $54 + 36 + 38 - 76$ **10.** $28 + 95 + 34 - 94$

11. $890 + 467 + 742 - 987$ **12.** $568 + 397 + 863 - 872$

13. $978 + 531 + 753 - 630$ **14.** $729 + 699 + 530 - 1,246$

15. $697 + 972 + 487 - 548$ **16.** $589 + 874 + 948 - 1,327$

2.7. Checking Subtraction. Since subtraction is the inverse operation of addition, the check of subtraction most used is the addition of the difference and the subtrahend to obtain the minuend. Thus, to check the subtraction shown at the left, we add
643   the difference 395 to the subtrahend 248 to obtain the
<u>248</u>   minuend 643. In other words, *we add up* as follows: 5
395   and 8 are 13, check 3 and carry 1; 9 and 1 and 4 are 14,
check 4 and carry 1; 3 and 1 and 2 are 6, check 6.

The method of checking by casting out 9's (see Section 1.8) can also be used, for evidently the sum of the check numbers of the difference and the subtrahend must equal the check number of the minuend. For example,

$$643 \quad 6 + 4 + 3 = 13 \text{ and } 1 + 3 = 4$$
$$\underline{248} \quad 2 + 4 + 8 = 14 \text{ and } 1 + 4 = 5$$
$$395 \quad 3 + 9 + 5 = 17 \text{ and } 1 + 7 = 8$$

The sum of the check numbers of the difference and the subtrahend is $8 + 5 = 13$ and $1 + 3 = 4$ which is equal to the check number of the minuend. Hence, the subtraction is correct.

Similarly, the method of casting out 11's (Section 1.8) can be used. Thus,

$$643 \quad (3 + 6) - 4 = 9 - 4 = 5$$
$$\underline{248} \quad (8 + 2) - 4 = 10 - 4 = 6$$
$$395 \quad (5 + 3) - 9 = 8 - 9 = (8 + 11) - 9 = 19 - 9 = 10$$

The sum of the check numbers of the difference and the subtrahend is $10 + 6 = 16$ and $6 - 1 = 5$ which is equal to the check number of the minuend. Hence, the subtraction is correct.

EXERCISE 7

Subtract and check by adding up.

1. 755	**2.** 968	**3.** 7,635	**4.** 85,892	**5.** 849,831
219	551	6,775	43,278	357,945
536	*417*	*860*	*42614*	*491886*

Subtract and check by (a) casting out 9's; (b) casting out 11's.

6. 376	**7.** 412	**8.** 9,401	**9.** 7,582	**10.** 6,785
178	234	5,382	3,846	3,529
198	*178*	*4019*	*3736*	*3256*

11. 8,915	**12.** 74,692	**13.** 94,251	**14.** 835,774	**15.** 653,682
3,796	57,534	45,817	545,968	374,199
5119	*17158*	*48434*	*289806*	*279683*

16. The readings of an electric meter, in kilowatt-hours, were as follows: Jul. 1, 7,241; Aug. 1, 7,488; Sept. 1, 7,742; Oct. 1, 8,032; Nov. 1, 8,358; Dec. 1, 8,708; Jan. 1, 9,080. (a) How many kilowatt-hours were consumed from July to January? (b) How many kilowatt-hours were consumed in each of the last six months?

17. The Republic of Colombia was founded by Simón Bolívar in 1819. It consisted of Venezuela, Colombia, Ecuador, and Panama, and had a total area of 996,385 square miles. After Venezuela seceded in 1829, Colombia was left with an area of 644,235 square miles. Ecuador seceded in 1830 and left Colombia with an area of 468,405 square miles. Finally, Panama seceded in 1903 and left Colombia with its present area of 439,830 square miles. (a) How many years after the foundation of the Republic of Colombia did each of the countries, Venezuela, Ecuador, and Panama, secede? (b) What is the area of each of the following countries: Venezuela, Ecuador, and Panama?

18. A man came to a store, bought a suit for $40, and gave the merchant a $100 bill. The merchant, not being able to change the bill, took it to a neighboring store and got it changed. After the man had left with his suit and change, the other store owner called and informed the merchant that the bill was counterfeit and that he was holding him responsible for the $100. After the merchant repaid the other store owner, what was his total loss?

19. If 9 birds eat 9 worms in 9 minutes, how many birds will it take to eat 100 worms in 100 minutes at the same rate?

20.   Johnson's cat went up a tree,
                        Which was sixty feet and three;
                        Every day he climbed eleven,
                        Every night he came down seven,
                        Tell me if he did not drop,
                        When his paws would touch the top.

Multiplication

3.1. Multiplication. The fundamental concepts of synthesis have been discussed in Chapter 1. The process of addition extends to numbers of any relative value, that is, it applies whether we add numbers with different relative values or equal numbers. Thus, a father of five, buying sneakers for his children at $4 a pair, can find the total cost by adding: $4 + $4 + $4 + $4 + $4 = $20; or he can note the fact that $4 is a repetition occurring 5 times, so that the total cost is 5 times $4. This process of uniting numbers that are all *equal*, with the idea of *times*, is called *multiplication*. Thus, multiplication is a special case of addition in which the numbers added are all equal.

Since a number is a set of units arranged in an ordered sequence (page 4), multiplication is the process of combining a sets, each of which contains b elements, into a single set C and determining the number property c of this larger set. The number b denoting the elements in each of the sets is called the *multiplicand*. The number a indicating the number of sets, each of which contains b elements, is the *multiplier*. Finally, c, denoting the number of elements in the resulting larger set C, is called the *product* and we write $a \times b = c$ (a times b is equal to c). In other words, the *product* of two numbers is the result obtained by taking one number as many times as there are units in the other. Multiplication can then be simply defined as the process of finding the product of two numbers. The number to be multiplied is the *multiplicand* and the number which multiplies is the *multiplier*. Taking $5 \times $4 = 20 as an example, $4 is the multiplicand; 5 is the multiplier; and $20 is the product.

3.2. Laws of Multiplication. The basic laws of multiplication are:

1) *The multiplier is always an abstract number.* This law follows

from the fact that the multiplier indicates the number of sets, each of which contains a given number of elements. In other words, the multiplier denotes the *number of times* the multiplicand is taken and evidently we cannot multiply pairs of shoes by dollars. Suppose the father of five buying sneakers at $4 a pair reasoned that 5 pairs of shoes times 4 dollars is 20 dollars and gave the salesman 20 dollars. His reasoning would be equally sound if he concluded that 5 pairs of shoes times 4 dollars is 20 pairs of shoes and insisted that he receive 20 pairs of shoes. In finding areas and volumes, the statements "feet times feet gives square feet" and "square feet times feet gives cubic feet" are made. These statements are misleading for the fact is that in finding the area of a rectangle for example, we multiply, *not* feet by feet which is absurd, but the number of square feet in the base by the number of such rows; and the number of rows is an abstract number.

2) *The product is a number of things similar to the things of the multiplicand.* This law is easily seen, for the product is the result of taking the multiplicand as many times as there are units in the multiplier. Or you may say, the product is the sum of the multiplicand used as many times as indicated by the multiplier, and, from the second law of addition (Section 1.3), the sum is a number of things similar to the things added. Thus, 5 times $4 is $20; 3 times 6 feet is 18 feet; and neither answer can be expressed in anything else but dollars or feet, respectively.

3) *The product is the same regardless of the order in which the numbers are multiplied;* thus, either number may be the multiplier leaving the other to be the multiplicand. This is known as the *commutative law* of multiplication and may be illustrated as follows. Consider a framework of beads and wires as shown in

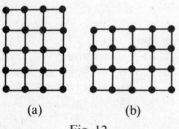

Fig. 12(a). There are 5 rows of 4 beads each, totaling 20 beads, so that 5 × 4 beads = 20 beads. Now suppose we turn this framework on its side, as shown in Fig. 12(b). We now have 4 rows of 5 beads each, and evidently the same total number of beads, that is, 4 × 5 beads = 20 beads. Therefore, 5 × 4 beads = 4 × 5 beads = 20 beads. If we let the beads represent elements of a set, Fig. 12(a)

(a)   (b)

Fig. 12.

can be thought of as illustrating 5 sets of 4 elements each, whereas
Fig. 12(b) illustrates 4 sets of 5 elements each and, as before, $4 \times 5 = 5 \times 4 = 20$. Thus, if a and b represent any integers, whether
we consider a sets of b elements each, or b sets of a elements each,
it is clear that the product will be the same. Hence,

$$a \times b = b \times a.$$

4) *To obtain the product of several numbers, any method of
grouping may be used.* This is known as the *associative law* of
multiplication and may be illustrated as follows. Figure 13 represents a framework of beads and wires which we can consider as
consisting of 5 tiers, each containing 3 rows of 4 beads. There
are $3 \times 4 = 12$ beads in each tier, and since there are 5 tiers, the
total number of beads is $5 \times 12 = 60$, so that $5 \times (3 \times 4) = 60$.
But by the commutative law, $5 \times (3 \times 4) = (3 \times 4) \times 5$; hence,
$(3 \times 4) \times 5 = 60$.

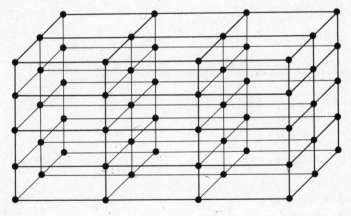

Fig. 13.

If we now consider the figure from the front to the back, we
can say that it consists of 3 rows, each containing 4 columns of
5 beads. There are now $4 \times 5 = 20$ beads in each row, and since
there are 3 rows, the total number of beads is $3 \times 20 = 60$.
Hence, $3 \times (4 \times 5) = 60$.

Moreover, we can consider the figure from right to left and say
that it consists of 4 columns, each containing 5 rows of 3 beads.
There are now $5 \times 3 = 15$ beads in each column, and since there

are 4 columns, the total number of beads is $4 \times 15 = 60$. Thus, $4 \times (5 \times 3) = 60$. Therefore,

$$(3 \times 4) \times 5 = 3 \times (4 \times 5) = 4 \times (5 \times 3) = 60.$$

This can be expressed by saying that if a, b, and c are any integers, then

$$a \times (b \times c) = (a \times b) \times c = (a \times c) \times b.$$

5) *The product of the sum of two or more numbers by a given multiplier is equal to the sum of the partial products obtained by multiplying each number by the given multiplier.* This law states that

$$(3 + 4) \times 2 = (3 \times 2) + (4 \times 2)$$
$$7 \times 2 = 6 + 8$$
$$14 = 14$$

and can be illustrated as follows. Consider the product $(3 + 4) \times 2$. The numbers 3 and 4 represent the two sets $\{\cdots\}$ and $\{\cdots\cdot\}$, respectively. The sum of these two sets is the set

$$\left\{ \underbrace{\cdots}_{3} \; \underbrace{\cdots\cdot}_{4} \right\}$$

Now if this resulting set is taken twice, the result

$$\left\{ \underbrace{\cdots\cdots\cdot}_{7} \; \underbrace{\cdots\cdots\cdot}_{7} \right\}$$

is a set containing 14 elements. But if the two sets $\{\cdots\}$ and $\{\cdots\cdot\}$ are each taken twice, we obtain $\left\{ \underbrace{\cdots}_{3} \; \underbrace{\cdots}_{3} \right\}\left\{ \underbrace{\cdots\cdot}_{4} \; \underbrace{\cdots\cdot}_{4} \right\}$ and their sum is a set $\left\{ \underbrace{\cdots\cdots}_{6} \; \underbrace{\cdots\cdots\cdot}_{8} \right\}$ also containing 14 elements. Thus, if a, b, and c are any integers,

$$a \times (b + c) = (a \times b) + (a \times c).$$

This is known as the *distributive law* of multiplication; it connects multiplication and addition. Thus, we say that multiplication is distributive with respect to addition, for it enables us to distribute the multiplication of the sum of two or more numbers over the numbers to be added.

The distributive law may also be illustrated by obtaining the areas of the rectangles shown in Fig. 14. The area of the large rectangle $MPQS$ is MP times MS, that is, $(b + c) \times a$. The

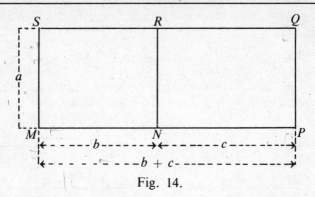

Fig. 14.

area of rectangle $MNRS$ is $b \times a$ and that of $NPQR$ is $c \times a$. Evidently, the area of the large rectangle $MPQS$ is equal to the sum of the areas of the rectangles $MNRS$ and $NPQR$. Therefore,

$$(b + c) \times a = (b \times a) + (c \times a).$$

3.3. Process of Multiplication. As in the case of addition, the fundamental tool for the operation of multiplication is the multiplication table based on the single-digit combinations. The usual steps in learning multiplication are:

1) The memorization of the single-digit combinations.

2) Multiplication of a number of more than one digit by a single digit, without carrying.

3) Multiplication of a number of more than one digit by a single digit, with carrying.

4) Multiplication of a number of more than one digit by two or more digits.

3.4. Single-Digit Combinations. There are ten digits. Each digit can be combined with any one of the ten digits to form a product. Hence, there are $10 \times 10 = 100$ single-digit combinations.

Of these one hundred combinations, the zero combinations can be eliminated on the following basis. Zero indicates an empty set, that is, a set that contains no elements. The product of zero by any other number indicates that the empty set is to be taken a certain number of times. No matter how many times the empty set is taken, the resulting set will still contain no elements. Hence, *zero multiplied by any number is zero.* That is, $N \times 0 = 0$. Moreover, to multiply any number by zero means that a given set is to be taken *no* times, that is, not at all. Hence, *any number multiplied by zero is zero.* Thus, $0 \times N = 0$ and we write

$$N \times 0 = 0 \times N = 0.$$

The multiplication table is shown in Fig. 15. The product of any single-digit combination can be found as follows. To find the product 3×4, run down the column headed m until you

MULTIPLICATION TABLE

m	1	2	3	4	5	6	7	8	9
1	1	2	3	4	5	6	7	8	9
2	2	4	6	8	10	12	14	16	18
3	3	6	9	12	15	18	21	24	27
4	4	8	12	16	20	24	28	32	36
5	5	10	15	20	25	30	35	40	45
6	6	12	18	24	30	36	42	48	54
7	7	14	21	28	35	42	49	56	63
8	8	16	24	32	40	48	56	64	72
9	9	18	27	36	45	54	63	72	81

Fig. 15.

reach 3; then move your finger along the horizontal row corresponding to 3 until you reach the column headed 4. The answer is found to be 12.

The importance of thoroughly memorizing these single-digit combinations until there is absolutely no hesitancy in producing the correct answer cannot be overemphasized. The answer should be arrived at instantly, without having to either count mentally or hesitate in any way.

Example. Find the product: (a) 2×5; (b) 3×7; (c) 7×6; (d) 8×7.

Solution. (a) 10. The product 10 should be seen immediately, without any hesitation, by simply looking at 2×5. (b) 21. (c) 42. (d) 56.

EXERCISE 8

At sight, give the product of each of the following.

1. 2×6	**6.** 8×2	**11.** 9×4	**16.** 8×8	**21.** 8×6
2. 4×3	**7.** 6×4	**12.** 5×9	**17.** 8×9	**22.** 3×4

3. 5 × 7 ³⁵   **8.** 4 × 5 ²⁰   **13.** 6 × 6 ³⁶   **18.** 7 × 4 ²⁸   **23.** 3 × 5 ¹⁵
4. 4 × 8 ³²   **9.** 7 × 3 ²¹   **14.** 7 × 7 ⁴⁹   **19.** 9 × 9 ⁸¹   **24.** 4 × 9 ³⁶
5. 5 × 3 ¹⁵   **10.** 3 × 2 ⁶   **15.** 6 × 7 ⁴²   **20.** 8 × 7 ⁵⁶   **25.** 6 × 9 ⁵⁴
26. How many times does a clock strike in a day if it strikes only the hours?
27. Five times what number is three times that number?

3.5. Multiplication of Numbers of Two or More Digits.

a) The multiplication of a number of two or more digits by a single-digit number in which the product of any one of the digits in the multiplicand by the single-digit multiplier is less than ten (that is, without carrying), is a simple application of the single-digit combinations and the laws of multiplication. For example, the operation 34 × 2 can be written (30 + 4) × 2 and, by the distributive law, is equal to (30 × 2) + (4 × 2) = 60 + 8 = 68. This operation can be arranged as follows.

$$34 = 3 \text{ tens} + 4 \text{ units}$$
$$\underline{\hspace{2cm} 2 \hspace{2cm}}$$
$$6 \text{ tens} + 8 \text{ units}$$

Using the principle of place value in our modern system of notation, the product is written 68. The operation is simply written as it is below.

EXPLANATION

34   4 × 2 = 8. Write 8 in the units place in the answer.
2   3 × 2 = 6. Write 6 in the tens place in the answer.
——
68

As in any arithmetic operation of this type, the answer should be seen immediately. It should not be necessary to say two fours are eight, and then two threes are six. Without any hesitation, one should think 8, 6.

b) Usually the multiplication of a number of two or more digits by a single-digit number involves carrying. That is, the product of any one of the digits in the multiplicand by the single-digit multiplier is more than ten. Again, the application of the appropriate laws of addition, the single-digit combinations, and the distributive law solves the problem easily. Thus, the operation 29 × 4 can be written (20 + 9) × 4 and, by the distributive law, is equal to (20 × 4) + (9 × 4) = 80 + 36 = 116. The operation can be arranged as follows.

$$29 = 2 \text{ tens} + 9 \text{ units}$$
$$\underline{\hspace{3cm} 4}$$
$$8 \text{ tens} + 36 \text{ units}$$

Now   8 tens = 8 tens

and   36 units = $\underline{3 \text{ tens} + 6 \text{ units}}$.   This sum is equal to

                   11 tens + 6 units

or   1 hundred + 1 ten + 6 units.

Using the principle of place value, the product is written 116.

In our modern system of notation, the operation is performed by multiplying the units digit first; reducing the resulting product to tens and units; writing the units in the units place in the answer (directly under the units column); carrying the tens part of this product to the tens column; adding it to the product of the tens digit in the multiplicand and the single-digit multiplier; and so on. Thus,

MENTAL CARRIES: (3)   EXPLANATION

         29       $9 \times 4 = 36$. Write 6 in the units place

         $\underline{4}$ and carry 3. Then, $2 \times 4 = 8$ and

      116      $8 + 3 = 11$. Write 1 in the tens place and 1 in the hundreds place.

c) The multiplication of a number of two or more digits by another number of two or more digits is an extension of the preceding process. For example, the operation 276×43 can be written $(200 + 70 + 6) \times (40 + 3)$. By the application of the distributive law then, we may multiply each term in the multiplicand by 3, find the product of each term in the multiplicand by 40, and finally, obtain the sum of the partial products. The whole process can be illustrated as follows.

$$
\begin{array}{r}
200 + 70 + 6 \\
\underline{40 + 3} \\
18 \ \dots \ (3 \times 6) \\
210 \ \dots \ (3 \times 70) \\
600 \ \dots \ (3 \times 200) \\
240 \ \dots \ (40 \times 6) \\
2{,}800 \ \dots \ (40 \times 70) \\
\underline{8{,}000 \ \dots \ (40 \times 200)} \\
11{,}868
\end{array}
$$

Note that if the operation is performed as shown, the order in which the numbers are multiplied is immaterial. That is, it makes no difference whether we multiply from right to left, left to right, or in any order whatever. Moreover, in multiplying by any whole number of tens, such as, 10, 20, 30, it is evident that the product will contain a zero in the units column; multiplying by any whole number of hundreds, 100, 200, 300, and so on, will yield a product with zeros in the units and the tens columns. For example, $40 \times 6 = 240$ and $400 \times 6 = 2,400$. Hence, we do not write these zeros and keeping place values in mind, the operation is reduced to single-digit combinations.

Ten Thousands	Thousands	Hundreds	Tens	Units
			1	8
		2	1	
		6		
		8	2	8
		2	4	
	2	8		
	8			
1	1	0	4	
1	1	8	6	8

$$\begin{array}{r} 276 \\ \times 43 \end{array}$$

$3 \times 6 \ldots (3 \times 6)$
$3 \times 7 \ldots (3 \times 70)$
$3 \times 2 \ldots (3 \times 200)$
First partial product
$4 \times 6 \ldots (40 \times 6)$
$4 \times 7 \ldots (40 \times 70)$
$4 \times 2 \ldots (40 \times 200)$
Second partial product
$828 + 11,040$

Fig. 16(a).

If the single-digit combinations are not written until the carries are mentally added, and we multiply from right to left keeping place values in mind, the operation can further be condensed as shown in Fig. 16 (b), page 48.

The whole operation can be written in a neat and concise manner by aligning the partial products as shown at the left. The operation 276×43 should be performed by thinking 3, 6, 18, write 8 carry 1; now 3, 7, 21, and 1 (carried), 22, write 2 carry 2; then 3, 2, 6, and 2 (carried), 8, write 8. Since the second partial product

$$\begin{array}{r} 276 \\ 43 \\ \hline 828 \\ 11\ 04 \\ \hline 11,868 \end{array}$$

is the result of multiplying by 40, we shift it one unit to the left
so that its single-digit combinations fall in the proper column.
Thus 4, 6, 24, write 4 (in the tens column under the 2 of the first
partial product) carry 2; now 4, 7, 28, and 2 (carried), 30, write 0
carry 3; then 4, 2, 8, and 3 (carried), 11. Finally, by adding the
partial products we obtain the answer, 11,868.

Ten Thousands	Thousands	Hundreds	Tens	Units		Write	Carry
				8	$3 \times 6 = 18$	8	1
			2		$(3 \times 7) + 1 = 22$	2	2
		8			$(3 \times 2) + 2 = 8$	8	
		8	2	8	First partial product		
			4		$(4 \times 6) = 24$	4	2
		0			$(4 \times 7) + 2 = 30$	0	3
1	1				$(4 \times 2) + 3 = 11$	1	1
1	1	0	4		Second partial product		
1	1	8	6	8	$828 + 11,040$		

$$276$$
$$\times 43$$

Fig. 16(b).

Note that we shift the second partial product one unit to the
left, the third partial product two units to the left, and so on
(see how the partial products, 4,368 and 2,912, are shown below,
at the left), to fix the place value of the single-digit combinations
mechanically. However, if their place values are kept in mind, the
order in which we obtain each partial product is immaterial. For
example,

```
    1,456              1,456              1,456
      234                234                234
    5 824              291 2              43 68
   43 68       or       43 68     or       5 824
  291 2                5 824              291 2
  340,704             340,704            340,704
```

Moreover, since any number multiplied by zero is zero, if 0 occurs in the multiplier, as in the case of 1,456 × 204, we could write a whole row of zeros as shown below, at the left.

1,456	1,456
204	204
5 824	5 824
00 00	291 2
291 2	297,024
297,024	

But this is a waste of time. So we simply skip the zero multiplier and multiply by the next number, 2 in this case, shifting its partial product two units to the left, as shown above at the right.

Example.   Multiply 64,204 by 2,306.
Solution.

EXPLANATION

64,204
2,306
385 224
19 261 2
128 408
148,054,424

Think 6, 4, 24, write 4 carry 2; 6, 0, 0 and 2, 2, write 2; 6, 2, 12, write 2 carry 1; 6, 4, 24 and 1, 25, write 5 carry 2; 6, 6, 36 and 2, 38. Skip the 0 multiplier. Then 3, 4, 12, write 2 in the hundreds column, carry 1; 3, 0, 0 and 1, 1, write 1; 3, 2, 6, write 6; 3, 4, 12, write 2 carry 1; 3, 6, 18 and 1, 19. Finally, 2, 4, 8, write 8 in the thousands column; 2, 0, 0, write 0; 2, 2, 4, write 4; 2, 4, 8, write 8; 2, 6, 12. The sum of the partial products is 148,054,424.

EXERCISE 9

Multiply at sight.

1. 34 × 2	**4.** 42 × 2	**7.** 42 × 3	**10.** 31 × 5
2. 23 × 3	**5.** 52 × 4	**8.** 64 × 2	**11.** 53 × 3
3. 12 × 3	**6.** 31 × 7	**9.** 31 × 6	**12.** 82 × 3

Multiply.

13. 23 × 4	**18.** 42 × 35	**23.** 495 × 38	**28.** 8,936 × 83
14. 75 × 8	**19.** 53 × 29	**24.** 824 × 53	**29.** 2,412 × 362
15. 54 × 4	**20.** 95 × 32	**25.** 527 × 68	**30.** 5,783 × 437
16. 37 × 9	**21.** 88 × 63	**26.** 7,509 × 29	**31.** 125,459 × 623
17. 28 × 12	**22.** 86 × 67	**27.** 4,146 × 74	**32.** 243,087 × 3,607

33. Mr. Nathan sold 14 acres of land at $50 an acre and received, as part payment, 80 acres of land worth $3 an acre. How much money did he receive in cash?

34. A plane leaves New York City at 7 A.M. for Miami and travels at the rate of 349 miles per hour. At the same time, another plane leaves

Miami for New York City and travels at the rate of 275 miles per hour. If the air distance from New York City to Miami is 1,106 miles, how far apart are these planes at 9 A.M.?

35. A cattleman bought 80 head of cattle at $40 a head. He sold 30 head at $45 each and 25 head at $48 each. How much must he get for the remaining cattle to make a profit of $400?

36. A water tank is filled by two pipes. The first one has a flow of 65 gallons per minute and the second, 32 gallons per minute. If the two pipes are opened, how many gallons will there be in the tank at the end of 1 hour and 58 minutes?

37. John has $80. Henry has twice as much as John, less $17; and Peter has as much as the other two together, plus $12. If they pool their money and spend $315 on a boat, how much will they have left?

3.6. Short Methods of Multiplication. A short method should abbreviate an operation considerably, or at least make the operation easier to perform. Some short methods that are worthwhile and easy to remember are illustrated here.

1) To multiply by 10, or any multiple of 10 by itself, such as 10, 100, 1,000: add as many zeros to the multiplicand as there are zeros in the multiplier.

Example. Multiply 528 by (a) 10; (b) 100; (c) 1,000.
Solution. (a) 5,280; (b) 52,800; (c) 528,000.

2) To multiply by any number whose value is within 5 units of any multiple of 10 by itself: multiply by the nearest multiple of 10 (as in the preceding example); then multiply by the units figure and add (or subtract) this result.

Example. Multiply: (a) 792 by 101; (b) 84 by 99; (c) 432 by 102.
Solution.

(a) 79,200
   792
79,992
EXPLANATION: 792 × 100 = 79,200.
79,200 + 792 = 79,992.

(b) 8,400
  84
8,316
EXPLANATION: 84 × 100 = 8,400.
8,400 − 84 = 8,316.

(c) 43,200
  864
44,064
EXPLANATION: 432 × 100 = 43,200.
432 × 2 = 864.
43,200 + 864 = 44,064.

3) To multiply by any number of two digits whose tens digit is one: multiply by the units digit; write this partial product one place to the right of the multiplicand; and add.

Example. Multiply 353 by 17.

Solution.   EXPLANATION: 353 × 7 = 2,471. Write this partial product
3 53   one place to the right of the multiplicand as shown at the left,
2,471   and add. This method saves the time it takes to write the
─────   multiplicand and the multiplier.
6,001

4) To multiply by any number of two digits whose unit digit is
one: multiply by the tens digit; write this partial product one
place to the left of the multiplicand; and add.

Example. Multiply 765 by 71.

Solution.   EXPLANATION: 765 × 7 = 5,355. Write this partial product
  765   one place to the left of the multiplicand, as shown at the left,
5,3 55   and add. Again, note that this method simply saves the time
─────   it takes to write the multiplicand and the multiplier.
5 4,315

5) To multiply a two-digit number by 11:

a) If the sum of the two digits in the multiplicand is less than
10, add the two digits. Then place this sum between the two
digits of the multiplicand.

Example. Multiply 54 by 11.

Solution. Here 5 + 4 = 9. Hence the answer is 594.

b) If the sum of the two digits in the multiplicand is ten or
more, add the two digits. Then place the units digit of this sum
between the two digits of the multiplicand and increase the tens
digit of the multiplicand by one.

Example. Multiply: (a) 68 by 11; (b) 64 by 11.

Solution. (a) Here 6 + 8 = 14. Place the 4 between 8 and 6. Add 1
to 6. The answer is 748. (b) Here 6 + 4 = 10. Place 0 between 4 and
6. Add 1 to 6. The answer is 704.

6) To multiply by 5, 25, 50, 250:

a) To multiply by 5, annex a zero to the multiplicand and
divide by 2.

b) To multiply by 25, annex two zeros to the multiplicand
and divide by 4.

c) To multiply by 50, annex two zeros to the multiplicand
and divide by 2.

d) To multiply by 250, annex three zeros to the multiplicand
and divide by 4.

Example. Multiply: (a) 129 by 5; (b) 128 by 25; (c) 150 by 50;
(d) 192 by 250.

Solution. (a) 1,290 ÷ 2 = 645; (b) 12,800 ÷ 4 = 3,200;
(c) 15,000 ÷ 2 = 7,500; (d) 192,000 ÷ 4 = 48,000.

EXERCISE 10

Using short methods perform the indicated operation.

1. 55 × 10	**8.** 333 × 12	**15.** 319 × 61	**22.** 263 × 103
2. 67 × 100	**9.** 726 × 12	**16.** 475 × 51	**23.** 236 × 99
3. 769 × 1,000	**10.** 639 × 15	**17.** 743 × 41	**24.** 891 × 98
4. 514 × 11	**11.** 479 × 18	**18.** 455 × 71	**25.** 482 × 5
5. 168 × 11	**12.** 166 × 19	**19.** 978 × 91	**26.** 549 × 25
6. 396 × 11	**13.** 892 × 17	**20.** 454 × 101	**27.** 185 × 50
7. 601 × 11	**14.** 267 × 31	**21.** 279 × 102	**28.** 148 × 250

3.7. Complementary Method of Multiplication. In order to promote the use of the Hindu-Arabic numerals, the medieval teacher did not require his students to learn the multiplication table beyond 5 × 10. To multiply under this limitation, various methods were developed. Of these, the best known is the *Complementary method.* It has endured to modern times because it lends itself to finger reckoning. To multiply any two digits, find the complements of these digits (Section 2.6). Multiply the complements. The units digit of this product is the units digit of the answer. Mentally, carry the tens digit of this product and add it to the difference between either one of the two given digits and the complement of the other. The result is the tens digit in the answer.

Example.   Multiply: (a) 7 × 6;   (b) 8 × 7.

Solution.   (a) The complement of 7 is 3 and the complement of 6 is 4.

$$\begin{array}{c} 7 \diagdown 3 \\ 6 \diagup \diagdown 4 \\ \hline 4 \quad 2 \end{array}$$
The product 3 × 4 = 12. Write 2, carry 1. Add 1 to the difference between 7 − 4 or 6 − 3. The sum, 4 (3 + 1), is the tens digit.

(b) The complement of 8 is 2 and the complement of 7 is 3.

$$\begin{array}{c} 8 \diagdown 2 \\ 7 \diagdown 3 \\ \hline 5 \quad 6 \end{array}$$
The product 2 × 3 = 6. Write 6, nothing to carry. The difference between 8 − 3 or 7 − 2 is 5, the tens digit of the answer.

As stated above, this method lends itself to finger reckoning.

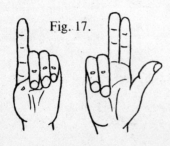
Fig. 17.

To obtain 9 × 7 find their respective complements, 1 and 3. Raise one finger on one hand and 3 fingers on the other hand to denote these complements. The sum of the closed fingers, 4 + 2 = 6, gives the tens digit; and the product of the raised fingers, 1 × 3 = 3, gives the units digit. The answer is 63.

If the multiplicand and the multiplier have the same number of digits and if their complements are small, this method can be extended to find the product of larger numbers. The product of the complements gives the units digit, or the units and tens digit, in the answer. The difference between any one of the given numbers and the complement of the other gives the digits at the left in the answer. Since the product must have twice as many digits as the multiplicand, we fill in the necessary number of zeros between the complementary product and the difference.

Example. Multiply: (a) 98 by 96; (b) 997 by 986.

Solution. (a) The complement of 98 is 2 and the complement of 96 is 4.

```
98 ⟍  2
96 ⟋  4
94    0 8
```

The product of these complements is $4 \times 2 = 8$. The difference $98 - 4 = 96 - 2 = 94$. Since the answer must have four digits, place one zero between 94 and 8, obtaining the answer 9,408.

```
997 ⟍  3
986 ⟋ 14
983   042
```

(b) The complement of 997 is 3 and the complement of 986 is 14. The product of these complements is $14 \times 3 = 42$. The difference $997 - 14 = 986 - 3 = 983$. Since the answer must have six digits, place one zero between 983 and 42, obtaining the answer 983,042.

3.8. Early Methods of Multiplication.

After trying and rejecting many methods, man, through mental effort and ingenuity, created our modern system of multiplication. Some of these early methods are illustrated here because their study can enrich your understanding of the principles used in our modern method.

Gelosia Method. The first printed case of this method is found in the Treviso arithmetic, printed in 1478.* The Italian word *gelosia* means the lattice work of a window. The method received this name because the calculations are noted in a framework resembling a window grating.

Figure 18 (a and b) shows the multiplication of 934 by 314 as it was illustrated in the Treviso arithmetic. For purposes of comparison, the operation performed in our modern system is shown at the left of the figure. The framework consists of a rectangle divided into cells by horizontal and vertical lines. The number of columns, 3, is the same as the number of digits in the multiplicand, 934. The number of rows, 3, is equal to the number of digits in the multiplier, 314. The multiplicand is written across the top, each digit heading a column.

---

*See David Eugene Smith, *History of Mathematics*, Vol. II (New York: Ginn, 1925).

If the multiplier is written down the right side, as shown in Fig. 18 (a), diagonals are drawn upward from left to right, dividing each cell into two parts. Each single-digit combination is

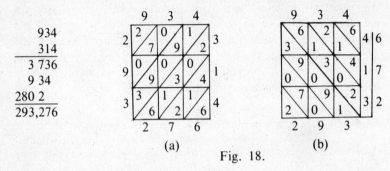

```
  934
  314
-------
 3 736
 9 34
-------
280 2
-------
293,276
```

Fig. 18.

written in a cell; the units digit in the lower portion and the tens digit, if any, in the upper portion. Thus, beginning at the upper right-hand corner, $3 \times 4 = 12$, write 2 in the lower portion and 1 in the upper portion. Then $3 \times 3 = 9$, write 9 in the lower portion of the next cell to the left and 0 in the upper portion; and so on. Finally, the numbers are added along the diagonal strips, beginning at the lower right-hand corner and carrying when necessary. Thus, 6; then $4 + 1 + 2 = 7$; then $2 + 0 + 3 + 1 + 6 = 12$, write 2, carry 1; then 1 (carried) $+ 1 + 9 + 0 + 9 + 3 = 23$, write 3, carry 2; then 2 (carried) $+ 0 + 7 + 0 = 9$, write 9, nothing to carry; and finally, 2. The answer is 293,276.

If the multiplier is written up the right side, as shown in Fig. 18 (b), diagonals are drawn downward from left to right, repeating the process explained in the preceding paragraph.

Note that the sum of the diagonals corresponds to the units column, tens column, and so on, in our modern method. Thus, the tens column as shown in our modern system (to the left of Fig. 18) is $3 + 4$ which is equal to the sum of the diagonal $4 + 1 + 2$ in Fig. 18(a), that is, $3 + 4 = 4 + 1 + 2 = 7$.

Example. Using the Gelosia method, multiply 346 by 78.

Solution. Beginning at the upper right-hand-corner cell, $7 \times 6 = 42$;

write 4 in the upper portion and 2 in the lower portion. Then $7 \times 4 = 28$; write 2 in the upper portion and 8 in the lower portion of the next cell to the left, and so on. Adding the diagonals, beginning at the lower right-hand corner, we obtain the answer 26,988.

Repiego Method. The "modo per repiego," that is, the method by decomposition, as illustrated by Pacioli in his *Suma* (1494), consists in expressing the multiplier in terms of its factors (see page 10) and then multiplying successively by each of the factors.

Example. Using the Repiego method, multiply 184 by 72.
Solution. $72 = 9 \times 8$. Then $184 \times 9 = 1,656$ and $1,656 \times 8 = 13,248$. *Ans.*

Scratch Method. Nicolo Tartaglia in his *General Trattato*, 1556, called this method "allo adietro" or the left-to-right method. The Arabs called it the *Hindu plan* and *Al-Nasavi*, one of their mathematicians, explained it in his works.[†] The order of the multiplication, as stated above, is always from left to right. The following example illustrates the successive steps.

Example. Using the Scratch method, multiply 376 by 46.
Solution. We first multiply by 4, that is, by 40; hence, write the numbers so that the units digit in the multiplicand is under the digit of highest

order in the multiplier. Thus, 376 is written under the 46 such that the 6 is under the 4 as shown above at the left. Beginning the multiplication with 4, we obtain $4 \times 3 = 12$. Write 2 over the 3 in the multiplicand and 1 to the left of 2. Scratch the 3 in 376 (see above left). Then $4 \times 7 = 28$. Write 8 over the 7 in the multiplicand and carry 2, adding it to the 2 above the 3, obtaining 4. Write 4 over the 2. Scratch this 2 and also the 7 in the multiplicand (see above center). Now $4 \times 6 = 24$. Write 4 over the 4 in the multiplier, that is, in 46. Carry 2, adding it to the 8 to obtain 10. Scratch the 8 and write 0 above it. Carry 1, adding it to the 4 above the 2, obtaining 5. Scratch the 4, write the 5 above this 4, and scratch the 6 in the multiplicand.

Rewrite the multiplicand one place to the right as shown at the left (page 55). Multiply 6 by 3, obtaining 18. Write 8 over 0, carry 1 and add it to the 5, obtaining 6. Write 6 over the 5. Scratch the 5, the 0 between the 4's, and the 3 in the multiplicand. Then $6 \times 7 = 42$ which when added to the 4 (over the 4 in the multiplier) gives 46. Write 6 over the 4 and scratch this 4. Carry 4, add it to the 8 to obtain 12. Write 2 over the 8. Scratch this 8, carry 1 and add it to the 6, obtaining 7. Scratch the 6 and write 7 over it. Scratch the 7 in the multiplicand. Finally, $6 \times 6 = 36$. Write 6 over the 6 in the multiplier. Carry 3 and add it to the 6 over the 4 to obtain 9. Scratch this 6 and write 9 over it. Scratch the 6 in the multiplicand. The answer, reading from left to right, is 17,296.

EXERCISE 11

Perform the indicated operation, using the Complementary method.

1. 8×9	**3.** 99×98	**5.** 95×91	**7.** 999×994
2. 9×5	**4.** 96×88	**6.** 998×996	**8.** 995×989

Perform the indicated operation using the Gelosia, Repiego, or Scratch method.

9. 321×27	**15.** $3,845 \times 324$	**21.** 357×36	**27.** 542×46
10. 548×67	**16.** $4,231 \times 408$	**22.** 457×35	**28.** 612×123
11. 281×36	**17.** 17×6	**23.** 182×27	**29.** 743×234
12. $2,758 \times 43$	**18.** 47×24	**24.** 421×21	**30.** 811×345
13. $3,914 \times 52$	**19.** 63×42	**25.** 283×25	**31.** $4,574 \times 213$
14. 457×231	**20.** 603×32	**26.** 487×37	**32.** $1,695 \times 251$

3.9. Checking Multiplication. Since multiplication is commutative, that is, the product is the same regardless of the order in which the numbers are multiplied, the operation can be checked by interchanging the multiplicand and the multiplier. Thus, the product 124×135 can be checked by finding the product 135×124 as shown below.

$$
\begin{array}{r}
124 \\
135 \\
\hline
620 \\
3\,72 \\
12\,4 \\
\hline
16,740
\end{array}
\qquad
\begin{array}{r}
135 \\
124 \\
\hline
540 \\
2\,70 \\
13\,5 \\
\hline
16,740
\end{array}
$$

The methods of casting out nines and casting out elevens (Section 1.8) are much quicker and easier. These methods can be used to check multiplication because the product of the check numbers

of the multiplicand and the multiplier is equal to the check number of the product.‡

Example.   Multiply 135 × 124 and check by (a) casting out 9's; (b) casting out 11's.

Solution.   (a) The check number of 135 is 1 + 3 + 5 = 9 or 0, and
135   the check number of 124 is 1 + 2 + 4 = 7.   The product
124   of the check numbers is 0 × 7 = 0. The check number of
540   16,740 = 1 + 6 + 7 + 4 + 0 = 18 = 1 + 8 = 9, or 0.
2 70   Since the product of the check numbers of the multiplicand
13 5   and the multiplier is equal to the check number of the prod-
16,740   uct, the answer is correct.

(b) The check number of 135 = (5 + 1) − 3 = 3. The check number of 124 = (4 + 1) − 2 = 3.   The product of the check numbers is 3 × 3 = 9. The check number of 16,740 = (0 + 7 + 1) − (4 + 6) = 8 − 10 or (8 + 11) − 10 = 19 − 10 = 9.   Since the product of the check numbers is equal to the check number of the product, the answer is correct.

EXERCISE 12

Multiply and check by (a) casting out nines; (b) casting out elevens.

1. 73 × 11   4. 479 × 19   7. 281 × 123   10. 3,214 × 275
2. 627 × 31   5. 98 × 88   8. 496 × 101   11. 3,082 × 203
3. 574 × 18   6. 996 × 997   9. 649 × 234   12. 4,605 × 607

13. If one pencil costs 6 cents, how much would 12 dozen pencils cost?
14. If each crate contains 17 dozen peaches, how many peaches are there in 14 crates?
15. How many minutes are there in the month of March?
16. If each page contains 42 lines, how many lines are there in a book of 238 pages?
17. The wheel of an automobile traveling at 60 miles per hour makes 672 revolutions in one minute. How many revolutions will it make in one hour?
18. If a capsule carrying an astronaut travels at the rate of 291 miles per minute, how far will it travel in 4 hours and 29 minutes?
19. A clerk divided a number by 7, instead of multiplying, obtaining 18 as a result. Find the correct answer.
20. Two horsemen, 16 miles apart, are traveling toward each other, each traveling at the rate of 4 miles per hour. A horsefly traveling at the rate of 10 miles an hour flies back and forth between the two horses until they meet. How far does the horsefly fly?

‡Proof of this statement is given on page 159.

Division

4.1. Division. Just as multiplication is a special case of addition in which the numbers added are all equal, division is a special case of subtraction in which the same number is successively subtracted. For example, if a man with $20 wants to buy shirts costing $5 each, he can find the number of shirts he can buy by repeatedly subtracting $5 until he has no money left, or until the amount he has left is not sufficient to buy another $5 shirt. Thus, $20 − $5 = $15; $15 − $5 = $10; $10 − $5 = $5; and $5 − $5 = 0. Since he has subtracted $5 four times, he can buy 4 shirts; but more important, he has discovered that $5 is contained four times in $20. The idea of division, which is implied in subtraction, intrinsically involves the idea of times that is inherent in the process of multiplication. Since 4 times $5 is $20, $20 may be considered as containing $5 four times. Therefore, *division* may be regarded as the process of finding the number of times that one number contains another and as such, it is the inverse operation of multiplication.

A number, as defined on page 4, is a set of units arranged in an ordered sequence. Division can then be defined as the process of separating a larger set C containing c elements into equal, smaller sets each containing b elements, and determining the number, a, of these smaller sets. The number c, denoting the number of elements in the larger set C, is called the *dividend*. The number b, designating the number of elements in each of the smaller sets, is called the *divisor*. The number a, indicating the number of smaller sets, is called the *quotient*. In other words, the quotient of two numbers is the number of times that the dividend contains the divisor. Briefly, the dividend is the number to be divided and the divisor is the number we divide by to obtain the quotient. For example, in the operation 20 ÷ 5 = 4, 20 is the dividend, 5 is the divisor, and 4 is the quotient.

4.2. Laws of Division. The process of division, as any mathematical operation, is governed by fundamental laws. The following are the most important.

1) *The things of the dividend and the divisor must always be the same.* This law follows immediately from the definition of division as repeated subtraction, for, in subtraction, the two terms must be similar. For example, we can investigate how many times 5 books are contained in 20 books but not, how many times 5 books are contained in 20 apples. Since there are no books contained in apples, the second consideration is absurd.

2) *The quotient is always an abstract number.* This law follows from the fact that division is the inverse operation of multiplication. Since the quotient indicates the *number of times* the dividend contains the divisor, the divisor, taken as many times as indicated by the quotient, must equal the dividend. Hence, the quotient can be regarded as a multiplier and, by the first law of multiplication, Section 3.2, must be abstract.

Division is not commutative for $6 \div 2$ is not equal to $2 \div 6$. Actually, the latter operation is impossible without the introduction of fractions. Furthermore, division is not associative for $(27 \div 9) \div 3 = 1$ is not the same as $27 \div (9 \div 3) = 9$. As in subtraction, the following two properties of division are sometimes erroneously confused with these laws.

3) *Either of the two smaller sets can be taken as the divisor and then the other will be the quotient.* Thus, if a, b, and c represent any integers and if $a \times b = c$, then $c \div a = b$ and $c \div b = a$. For example, if $2 \times 3 = 6$, then $6 \div 2 = 3$ and $6 \div 3 = 2$.

4) *The result of dividing a given number by a second number and then dividing the quotient thus obtained by a third number is the same as the result obtained by dividing the given number by the third number and then dividing the quotient thus obtained by the second number.* Thus if a, b, and c are any integers, then $a \div b \div c = a \div c \div b$. For example, $27 \div 9 \div 3 = 27 \div 3 \div 9 = 1$. Note that this is not the same as the associative law.

5) *The quotient obtained by dividing the sum of two or more numbers by a given divisor is equal to the sum of the partial quotients obtained by dividing each number by the given divisor.* In general, this law states that if a, b, and c are any integers, then

$$(a + b) \div c = (a \div c) + (b \div c).$$

In particular,

$$(10 + 6) \div 2 = (10 \div 2) + (6 \div 2)$$
$$16 \div 2 = 5 + 3$$
$$8 = 8.$$

This is known as the *distributive law* of division; it relates division with addition. Thus we say that division is distributive with respect to addition, for it enables us to distribute the division of the sum of two or more numbers over the numbers to be added.

6) *Division by zero is excluded.* As previously stated, division is the inverse operation of multiplication. Thus, $6 \div 2 = 3$ if and only if $6 = 2 \times 3$. Consider the division of a number (say 5) by zero. If we let $5 \div 0 = 0$, then $5 = 0 \times 0$. But $0 \times 0 = 0$, hence this possibility is absurd. Moreover, if we let $5 \div 0 = 5$, then $5 = 0 \times 5$. But $0 \times 5 = 0$, making this possibility also absurd.

In general, if a, b, and c are any integers, to divide a by b is to find a unique number c such that $a = b \times c$. That is, $a \div b = c$ if and only if $a = b \times c$. If we consider $b = 0$, two cases arise.

a) If $a \neq 0$ (that is, a is not equal to zero), to divide a by zero is to find a unique number c such that $a = 0 \times c$ where $a \neq 0$. But this is impossible, for $0 \times c = 0$ for any number c.

b) If $a = 0$, to divide a by zero is to find a unique number c such that $a = 0 \times c$ or in this case such that $0 = 0 \times c$. But $0 \times c = 0$ for any number c. Therefore, the number c cannot be determined uniquely.

It follows that if the division of one number by another is to yield a unique quotient, division by zero must be excluded.

7) *The dividend is always equal to the product of the divisor and the quotient, plus the remainder.* If any integer is divided by another smaller integer, greater than zero, then either the divisor is contained in the dividend an exact number of times, or it lies between two successive multiples of the divisor. In the first case, the remainder is zero and in the second case, the remainder is a number greater than zero and less than the divisor. In any case,

$$\text{dividend} = \text{divisor} \times \text{quotient} + \text{remainder}.$$

This is simply an explicit statement of the general definition, in order to cover the case when the divisor is not contained an integral number of times in the dividend. Thus, since $2 \times 3 = 6$

and $2 \times 4 = 8$, then $6 \div 2 = 3$ and $8 \div 2 = 4$. Now 7 lies between 6 and 8; hence, 2 is contained in 7 three times with a remainder left over. The operation $7 \div 2$ can be expressed in either of the following ways.*

$$
\begin{array}{c}
3 \\
2\overline{)7} \\
6 \\
\hline
1
\end{array}
\qquad \text{or} \qquad
\begin{array}{c}
3\tfrac{1}{2} \\
2\overline{)7} \\
7 \\
\hline
0
\end{array}
$$

At the right, above, the division is expressed with a remainder of zero and simply states that $7 \div 2 = 3\tfrac{1}{2}$ where $3\tfrac{1}{2}$ is the quotient. This notation is clumsy and in most cases unnecessary. At the left, the division is expressed with a remainder. Here 7 is the dividend, 2 is the divisor, 3 is the quotient, and 1 is the remainder, so that $7 = 2 \times 3 + 1$. If the divisor is contained an integral number of times in the dividend, such as $6 \div 2 = 3$, the remainder is 0 and we can write $6 = 2 \times 3 + 0$. Thus there is no exception to the statement—dividend = divisor × quotient + remainder.

8) *Multiplying or dividing both divisor and dividend by the same number, except zero, does not change the quotient.* For example, $6 \div 2 = 3$; $(6 \times 5) \div (2 \times 5) = 30 \div 10 = 3$ and $(6 \div 2) \div (2 \div 2) = 3 \div 1 = 3$. The statement also holds for $7 \div 2 = 3\tfrac{1}{2}$, where the remainder is not zero. Thus, $(7 \times 3) \div (2 \times 3) = 21 \div 6 = 3\tfrac{1}{2}$ and $(7 \div 2) \div (2 \div 2) = 3\tfrac{1}{2} \div 1 = 3\tfrac{1}{2}$.†

BASIC DIVISION COMBINATIONS

$4 \div 2 = 2$	$6 \div 2 = 3$	$8 \div 2 = 4$	$10 \div 2 = 5$
$6 \div 3 = 2$	$9 \div 3 = 3$	$12 \div 3 = 4$	$15 \div 3 = 5$
$8 \div 4 = 2$	$12 \div 4 = 3$	$16 \div 4 = 4$	$20 \div 4 = 5$
$10 \div 5 = 2$	$15 \div 5 = 3$	$20 \div 5 = 4$	$25 \div 5 = 5$
$12 \div 6 = 2$	$18 \div 6 = 3$	$24 \div 6 = 4$	$30 \div 6 = 5$
$14 \div 7 = 2$	$21 \div 7 = 3$	$28 \div 7 = 4$	$35 \div 7 = 5$
$16 \div 8 = 2$	$24 \div 8 = 3$	$32 \div 8 = 4$	$40 \div 8 = 5$
$18 \div 9 = 2$	$27 \div 9 = 3$	$36 \div 9 = 4$	$45 \div 9 = 5$

*For the definition of a fraction, see Chapter 6.
†See the rules for operations with fractions given in Chapter 6.

4.3. Types of Work in Division.　The steps used to introduce the process of division are:

1) Division by a single-digit number with no remainder and without carrying.

2) Division by a single-digit number with a remainder but without carrying.

3) Division by a single-digit number with carrying.

4) Division by a number of two or more digits.

4.4. Single-Digit Divisors.

1) To divide by a single-digit number with no remainder and without carrying, first memorize the basic division combinations. These combinations can be obtained by reversing the process of multiplication on the Multiplication Table (Fig. 15). Thus considered, the body of the table contains the dividends; the first row at the top, the divisors; and the first column at the left, the quotients. To obtain $72 \div 8$, locate the column headed 8; run down this column until you find the dividend 72; then move your finger to the left, along this horizontal row, until you reach the column headed m. The quotient is 9. However, because there should be no hesitancy in producing the correct answer, these combinations are given as division operations in Fig. 19. Since zero divided by any number except zero is equal to zero, and any number divided by itself is equal to 1 ($1 = 1 \div 1 = 2 \div 2 = 3 \div 3$, and so on), these combinations are omitted. Thus, Fig. 19 gives the seventy-two basic combinations of division.

BASIC DIVISION COMBINATIONS

$12 \div 2 = 6$	$14 \div 2 = 7$	$16 \div 2 = 8$	$18 \div 2 = 9$
$18 \div 3 = 6$	$21 \div 3 = 7$	$24 \div 3 = 8$	$27 \div 3 = 9$
$24 \div 4 = 6$	$28 \div 4 = 7$	$32 \div 4 = 8$	$36 \div 4 = 9$
$30 \div 5 = 6$	$35 \div 5 = 7$	$40 \div 5 = 8$	$45 \div 5 = 9$
$36 \div 6 = 6$	$42 \div 6 = 7$	$48 \div 6 = 8$	$54 \div 6 = 9$
$42 \div 7 = 6$	$49 \div 7 = 7$	$56 \div 7 = 8$	$63 \div 7 = 9$
$48 \div 8 = 6$	$56 \div 8 = 7$	$64 \div 8 = 8$	$72 \div 8 = 9$
$54 \div 9 = 6$	$63 \div 9 = 7$	$72 \div 9 = 8$	$81 \div 9 = 9$

Fig. 19.

The application of the distributive law of division reduces the division of a number of two or more digits by a single-digit number, without carrying and without a remainder, to the basic combinations. For example, $684 \div 2 = (600 + 80 + 4) \div 2$ and, applying the distributive law, this is equal to $(600 \div 2) + (80 \div 2) + (4 \div 2) = 300 + 40 + 2 = 342$. This operation can be arranged as follows.

$$
\begin{array}{r}
300 + 40 + 2 \\
\hline
2)\overline{600 + 80 + 4} \\
600 \\
\hline
0 + 80 \\
80 \\
\hline
0 + 4 \\
4 \\
\hline
0
\end{array}
\qquad \text{or} \qquad
\begin{array}{r}
300 + 40 + 2 \\
\hline
2)\overline{600 + 80 + 4} \\
4 \\
\hline
80 + 0 \\
80 \\
\hline
600 + 0 \\
600 \\
\hline
0
\end{array}
$$

Note that if the divisor is contained in each unit of the dividend an integral number of times, the order in which the operation is performed is immaterial. Thus, at the left, above, the division is performed from left to right while at the right, it is performed from right to left. However, when the divisor is not contained in each unit of the dividend an integral number of times, it is more convenient to divide from left to right. Moreover, keeping place values in mind, the operation can then be written as shown at the left, below. This may be further condensed by not writing the zeros, as shown in the center, and finally, as shown at the right, by carrying out the operation mentally.

$$
\begin{array}{r}
342 \\
\hline
2)\overline{684} \\
6 \\
\hline
08 \\
8 \\
\hline
04 \\
4 \\
\hline
0
\end{array}
\qquad \text{or} \qquad
\begin{array}{r}
342 \\
\hline
2)\overline{684} \\
6 \\
\hline
8 \\
8 \\
\hline
4 \\
4 \\
\hline
0
\end{array}
\qquad \text{or} \qquad
\begin{array}{r}
342 \\
\hline
2)\overline{684}
\end{array}
$$

2) Division by a single-digit number with a remainder but without carrying involves finding the greatest integer which when multiplied by the divisor will yield a product less than, but nearest to the dividend. The difference between this product and the dividend is the remainder.

Example. Divide 17 by 5.

Solution. The greatest integer which when multiplied by 5 will yield a product less than, but nearest to 17, is 3. The product is $3 \times 5 = 15$ and the difference is $17 - 15 = 2$. Hence, 5 goes into 17 three times with a remainder of 2, that is,

$$17 = 5 \times 3 + 2$$

$$\begin{array}{r} 3 \\ 5\overline{)17} \\ 15 \\ \hline 2 \end{array}$$

Example. Divide 849 by 4.

Solution. Here, 4 goes into 8 twice. Since 8 denotes the hundreds digit in the dividend, the operation actually performed is $800 \div 4$; hence, write 2 over the 8 to represent the hundreds digit in the quotient. Then $4 \times 2 = 8$; write 8 under the 8 in the dividend and subtract, obtaining 0. Bring down the 4, then 4 goes into 4 once. Since 4 denotes the tens digit in the dividend, the operation actually performed is $40 \div 4$; hence, write 1 over the 4 in the dividend to represent the tens digit in the quotient. Then $4 \times 1 = 4$; write 4 under the 4 brought down and subtract, obtaining 0. Bring down the 9. Now the greatest integer which, when multiplied by 4, will yield a product less than, but nearest to 9, is 2. Hence, write 2 over the 9 in the dividend to denote the units digit in the quotient. Then, $4 \times 2 = 8$; write 8 under the 9 brought down and subtract, obtaining a remainder of 1. Here, 849 is the dividend, 4 is the divisor, 212 is the quotient, and 1 is the remainder. Thus, $849 = 4 \times 212 + 1$. The whole operation should be performed mentally. Thus, 4 goes into 8 twice with no remainder. Write 2 over the 8. Then $4 \div 4 = 1$ with no remainder, so write 1 over the 4. Finally, 4 goes into 9 twice with a remainder of 1.

$$\begin{array}{r} 212 \\ 4\overline{)849} \\ 8 \\ \hline 4 \\ 4 \\ \hline 9 \\ 8 \\ \hline 1 \end{array}$$

$$\begin{array}{r} 212 \\ 4\overline{)849} \\ \hline 1 \end{array}$$

3) The operation, division by a single-digit number with carrying is an application of the first two steps explained in the preceding paragraphs. If the single-digit divisor is not contained an integral number of times in any one given digit of the dividend, find the greatest integer which when multiplied by the divisor will yield a product less than, but nearest to the given digit in the dividend. This product of the greatest integer and the divisor is then subtracted from the given digit in the dividend. Finally, the remainder is expressed in terms of the next lower unit in the dividend. Now the reason why we divide from left to right is obvious. For example, in the operation $75 \div 3$, if we were to divide from right to left, 3 goes into 5 once with 2 left over. This remainder 2 could not be expressed easily in terms of the tens unit. However, if the tens digit, 7, is divided first, then 3 is contained in 7 twice with

a remainder of 1. This remainder of 1 ten can be easily combined with the units digit, 5, to give 15 units and $15 \div 3 = 5$. Consequently, we divide from left to right and arrange the work as shown in the following example.

Example. Divide 75 by 3.
Solution. Starting at the left in the dividend, the greatest integer which, when multiplied by 3, will yield a product less than, but nearest to 7, is 2. Write 2 above the 7 in the quotient. Then $3 \times 2 = 6$; write 6 under the 7 and subtract, obtaining 1. Bring down the 5 from the dividend and write it to the right of the remainder, 1. Since this remainder, 1, represents tens, 15 correctly forms the number fifteen. Now $15 \div 3 = 5$. Write 5 above the 5 in the dividend; multiply $3 \times 5 = 15$; write 15 under the 15 and subtract, obtaining 0. The answer is 25. This process should be accomplished mentally as far as possible. Thus, we should think 3 is contained in 7 twice with a remainder of 1. Write 1 under the 7 in the dividend and bring down the 5. Finally, $15 \div 3 = 5$. Write 5 over the 5 in the dividend. The answer is 25.

$$\begin{array}{r} 25 \\ 3\overline{)75} \\ 6 \\ \hline 15 \\ 15 \\ \hline 0 \end{array}$$

$$\begin{array}{r} 25 \\ 3\overline{)75} \\ 15 \end{array}$$

Example. Divide 9,732 by 5.
Solution. Starting at the left in the dividend, 5 goes into 9 once. Write 1 over the 9 in the dividend. Then $9 - (5 \times 1) = 4$. Write 4 under the 9 in the dividend and bring down the 7. Now 5 goes into 47 nine times. Write 9 over 7 in the dividend. Then $47 - (5 \times 9) = 2$. Write 2 under the 7 in 47 and bring down the 3 from the dividend. Then 5 goes into 23 four times. Write 4 over the 3 in the dividend. Next $23 - (5 \times 4) = 3$. Write 3 under the 3 in 23 and bring down the 2 from the dividend. Then 5 goes into 32 six times. Write 6 above the 2 in the dividend. After that, $32 - (5 \times 6) = 2$; hence, write 2 under the 2 in 32. It follows that $9,732 = 5 \times 1,946 + 2$.

$$\begin{array}{r} 1,946 \\ 5\overline{)9,732} \\ 4\,7 \\ \hline 23 \\ \hline 32 \\ \hline 2 \end{array}$$

EXERCISE 13

Divide at sight.

1. $12 \div 6$	**10.** $81 \div 9$	**19.** $72 \div 9$	**28.** $18 \div 3$
2. $45 \div 5$	**11.** $18 \div 9$	**20.** $21 \div 7$	**29.** $54 \div 9$
3. $4 \div 2$	**12.** $24 \div 8$	**21.** $24 \div 6$	**30.** $40 \div 8$
4. $9 \div 3$	**13.** $28 \div 7$	**22.** $56 \div 7$	**31.** $64 \div 2$
5. $16 \div 4$	**14.** $30 \div 6$	**23.** $27 \div 9$	**32.** $93 \div 3$
6. $25 \div 5$	**15.** $12 \div 4$	**24.** $32 \div 4$	**33.** $48 \div 2$
7. $36 \div 6$	**16.** $20 \div 5$	**25.** $35 \div 7$	**34.** $648 \div 2$
8. $49 \div 7$	**17.** $42 \div 7$	**26.** $63 \div 9$	**35.** $396 \div 3$
9. $64 \div 8$	**18.** $56 \div 8$	**27.** $48 \div 8$	**36.** $59 \div 8$

Perform the indicated operation.

37. 61 ÷ 9	**41.** 286 ÷ 4	**45.** 9,675 ÷ 3	**49.** 24,169 ÷ 7
38. 107 ÷ 5	**42.** 9,634 ÷ 3	**46.** 2,575 ÷ 5	**50.** 23,130 ÷ 4
39. 219 ÷ 7	**43.** 16,847 ÷ 4	**47.** 8,527 ÷ 2	**51.** 35,378 ÷ 6
40. 367 ÷ 9	**44.** 4,979 ÷ 7	**48.** 969 ÷ 7	**52.** 31,255 ÷ 9

53. How much time will a bricklayer need to do 560 square feet of wall if he works 8 hours a day and does 5 square feet per hour?

54. A water tank is filled by two pipes. The first one has a flow of 130 gallons every 2 minutes, and the second, 160 gallons every 5 minutes. How long will it take to fill the tank, if the capacity of the tank is 11,640 gallons?

55. Mr. Mead sold 12 acres of land at $500 an acre. With the income from this sale he bought 8 acres of land. How much did he pay per acre?

4.5. Divisors of Two or More Digits. Of all the elementary operations of arithmetic, division was considered the most difficult. Pacioli (1494) remarked that "If a man can divide well, everything else is easy, for all the rest is involved therein," and Hylles (1600) wrote, "Diuision is esteemed one of the busiest operations of Arithmetick, and such as requireth a mynde not wandering, or setled vppon other matters."[1] Actually, our modern system of notation has eliminated almost all of the original difficulties. The only problem that may arise is the estimating of the successive digits in the quotient. Many rules have been devised to help estimate these quotient-digits. The fact is that numerical judgment should be developed by constant practice so that no rules are necessary. Thus, in ascertaining how many times 21 goes into 69, the answer 3 and the remainder 6 should be seen immediately, after having performed the complete operation mentally. The part that the other digits in the divisor play in determining the quotient should be emphasized. A divisor of 21 is 2 tens and 1 unit or practically 2 tens, whereas a divisor of 28 is 2 tens and 8 units, that is, almost 3 tens. Thus, if 69 is to be divided by 28, the quotient is probably the same as that of 7 ÷ 3. In fact, $2 \times 28 = 56$ and $69 - 56 = 13$, which is the remainder. Both of these operations are shown below.

$$
\begin{array}{r} 3 \\ 21\overline{)69} \\ 63 \\ \hline 6 \end{array}
\qquad\qquad
\begin{array}{r} 2 \\ 28\overline{)69} \\ 56 \\ \hline 13 \end{array}
$$

[1] Smith, *op. cit.*

Division by a number of two or more digits is an extension of the preceding cases. For example, the operation $6{,}754 \div 21$ can be written $(6000 + 700 + 54 + 4) \div 21$. Then, applying the distributive law of division, each term can be divided by 21, in which case, the result of the complete operation can be obtained by adding the partial quotients thus obtained. The whole process can be illustrated as follows.

$$
\begin{array}{r}
300 + 20 + 1 \\
20 + 1\overline{)6000 + 700 + 50 + 4} \\
\underline{6000 + 300} \ldots \ldots \ldots \ldots \ldots (20 + 1) \times 300 \\
400 + 50 \\
\underline{400 + 20} \ldots \ldots \ldots \ldots (20 + 1) \times 20 \\
30 + 4 \\
\underline{20 + 1} \ldots \ldots \ldots (20 + 1) \times 1 \\
10 + 3
\end{array}
$$

Starting with the digit of highest order in the dividend, that is, with the digit at the left, $6000 \div 20 = 300$. Write 300 over the dividend, multiply $(20 + 1) \times 300$, and write the product $6000 + 300$ under the proper numbers in the dividend. Subtracting, we obtain a remainder of 400. Then $400 \div 20 = 20$, so add 20 to the quotient, now the quotient, thus far, shows $300 + 20$. Now multiply $(20 + 1) \times 20$ and write the product $400 + 20$ under the proper numbers. Subtracting, we obtain a remainder of 30. Next, 20 goes into 30 once; so, add 1 to the quotient and multiply $(20 + 1) \times 1$. Write the product $20 + 1$ under $30 + 4$ and subtract, obtaining the remainder $10 + 3$. Thus, the operation $6{,}754 \div 21$ has a quotient of 321 and a remainder of 13.

Note that if the dividend and the divisor are restored to their original form, the operation proceeds in the same manner, as shown at the left. Moreover, in determining the first digit in the quotient, one does not have to think $6000 \div 20$ but merely $6 \div 2 = 3$. Write 3 over the 7 in the dividend, multiply 21×3 and write the product 63 under the first two figures in the dividend. Subtract, obtaining a remainder of 4. Bring down the 5 from the dividend. Then $4 \div 2 = 2$; write 2 over the 5 in the dividend; multiply 21×2 and write the product 42 under the 45. Subtract, obtaining a remainder of 3. Bring down the 4 from the dividend. Since 2 goes into 3 once, write 1 over the 4 in the divi-

$$
\begin{array}{r}
321 \\
21\overline{)6{,}754} \\
\underline{6\ 3} \\
45 \\
\underline{42} \\
34 \\
\underline{21} \\
13
\end{array}
$$

dend; multiply 21 × 1 and write the product 21 under the 34. Subtract, obtaining a remainder of 13.

Example. Divide 97,182 by 47.

Solution. Starting at the left in the dividend, consider how many times 4 goes into 9 and use the answer, 2, as the first digit in the quotient.

```
        2,067
    47) 97,182
        94
         3 18
         2 82
           362
           329
            33
```

Write 2 over the 7 in the dividend; multiply 47 × 2 and write the product 94 under 97. Subtract, obtaining a remainder of 3. Bring down the 1 from the dividend. Since 47 does not go into 31, write a 0 in the quotient over the 1 in the dividend and bring down the 8 from the dividend. Now to find out how many times 47 goes into 318, consider how many times 4 goes into 31; the answer is 7. But 47 × 7 = 329, which is larger than 318; hence, we cannot use 7. We must use a smaller number, so we try 6. In fact, since 47 is nearer to 50 than it is to 40, we could have considered how many times 5 goes into 31, obtaining the correct digit, 6.[§] Write 6 over the 8 in the dividend; multiply 47 × 6 and write the product 282 under the 318. Subtract, obtaining a remainder of 36. Bring down the 2 from the dividend. Now 4 goes into 36 nine times. But evidently 9 is too large because of the influence of the other digit 7 in the divisor, since 47 × 9 = 423. Hence, we again determine how many times 5 goes into 36, and write 7 as the next digit in the quotient over the 2 in the dividend. Multiply 47 × 7 and write the product 329 under 362. Finally, subtract 362 − 329, obtaining the remainder 33.

Example. Divide 340,996 by 148.

Solution. Here the first digit in the divisor is 1. If we were to consider this digit only, in obtaining the digits in the quotient, we would probably use, as trial-quotients, digits that would be too

```
         2304
    148) 340996
         296
          449
          444
           596
           592
             4
```

large. Starting at the left in the dividend, 3 ÷ 1 = 3, but if 3 is used as a trial-quotient, the 4 in the tens digit of the divisor will cause the product of 148 × 3 to be in the 4 hundreds. Hence, we consider the first two digits in the divisor and estimate how many times 14 will be contained in 34, the first two digits in the dividend. The answer is 2, for 14 × 2 = 28, whereas 14 × 3 = 42 is too large. Write 2 over the 0 in the dividend; multiply 148 × 2 and subtract this product, 296, from 340 to get a remainder of 44. Bring down the 9 from the dividend. Then 14 goes

[§] Note that if we had tried this approach with the first digit, 5 goes into 9 once, then 47 × 1 = 47 and 97 − 47 = 50, which leaves a remainder that is too large. Thus, one must use one's own numerical judgment to determine, with the least effort, which digit should be used.

into 44 three times. Write 3 as the next digit in the quotient; multiply 148 × 3 and subtract this product, 444, from 449 to obtain a remainder of 5. Bring down the other 9 from the dividend. Obviously, 148 does not go into 59; hence, write 0 as the next digit in the quotient and bring down the 6 from the dividend. Now 14 goes into 59 four times; so, write 4 as the next digit in the quotient; multiply 148 × 4 and subtract this product, 592, from 596 to obtain a remainder of 4.

Example. A department store bought 45 alarm clocks at $5 each. They sold 15 of these clocks at $8 each. At what price must they sell the remaining clocks to make a total profit of $105?
Solution.

$$\text{Total cost} = 45 \times \$5 = \$225.$$
$$\text{Total sales price} = \$225 + \$105 = \$330.$$
$$\text{They sold } 15 \times \$8 = \underline{\$120.}$$
$$\text{Balance to be obtained} = \$210.$$

Remaining clocks = 45 − 15 = 30.
Sales price per clock = $210 ÷ 30 = $7. *Ans.*

Example. Moran & Co. bought 23 blankets and 10 wool jackets for $218. Later, at the same prices, they bought 15 blankets and 20 wool jackets for $250. How much did they pay for: (a) a blanket? (b) a jacket?
Solution. (a) If 23 blankets + 10 jackets cost $218, then twice the amount of merchandise will cost twice as much, that is,

$$46 \text{ blankets} + 20 \text{ jackets} = \$436.$$
$$\text{But} \quad \underline{15 \text{ blankets} + 20 \text{ jackets} = \$250.}$$
$$\text{Hence, } 31 \text{ blankets} \qquad\qquad\ = \$186.$$

Therefore, 1 blanket = $186 ÷ 31 = $6. *Ans.*

$$\text{(b) Since } 23 \text{ blankets} + 10 \text{ jackets} = \$218,$$
$$\text{then } 23 \times \$6 + 10 \text{ jackets} = \$218,$$
$$\text{and } 10 \text{ jackets} = \$218 - \$138 = \$80.$$

Therefore, 1 jacket = $80 ÷ 10 = $8. *Ans.*

EXERCISE 14

Divide at sight.

1. 24 ÷ 12	**6.** 76 ÷ 19	**11.** 85 ÷ 17	**16.** 144 ÷ 16
2. 56 ÷ 14	**7.** 70 ÷ 14	**12.** 126 ÷ 18	**17.** 95 ÷ 19
3. 96 ÷ 16	**8.** 105 ÷ 21	**13.** 60 ÷ 15	**18.** 78 ÷ 13
4. 65 ÷ 13	**9.** 64 ÷ 16	**14.** 91 ÷ 13	**19.** 117 ÷ 13
5. 42 ÷ 14	**10.** 102 ÷ 17	**15.** 136 ÷ 17	**20.** 152 ÷ 19

Divide.

21. 952 ÷ 17	**32.** 37,291 ÷ 74
22. 644 ÷ 23	**33.** 48,012 ÷ 68
23. 735 ÷ 21	**34.** 22,105 ÷ 144
24. 888 ÷ 37	**35.** 97,640 ÷ 904
25. 952 ÷ 28	**36.** 207,019 ÷ 659
26. 4,368 ÷ 56	**37.** 118,759 ÷ 576
27. 5,185 ÷ 17	**38.** 1,598,000 ÷ 782
28. 7,326 ÷ 18	**39.** 6,749,182 ÷ 3,289
29. 7,107 ÷ 23	**40.** 7,074,094 ÷ 2,017
30. 16,059 ÷ 53	**41.** 9,426,000 ÷ 8,291
31. 21,408 ÷ 96	**42.** 14,849,999 ÷ 6,172

43. A merchant bought 50 suits at $25 each. He sold 20 suits at $35 each. At what price must he sell each of the remaining suits to make a total profit of $350?

44. The owner of a haberdashery bought 12 hats and 25 pairs of shoes for $235. Later he bought, at the same price, 36 hats and 6 pairs of shoes for $222. How much did he pay for: (a) a pair of shoes? (b) a hat?

45. An airplane leaves a city at 9A.M. traveling at the rate of 180 miles per hour. A second plane leaves the same city at 10 A.M. to overtake the first and travels at the rate of 240 miles per hour. At what time will the second plane overtake the first?

46. If 2 dogs and 5 puppies weigh 63 pounds and 5 dogs and 2 puppies weigh 126 pounds, how much does: (a) a puppy weigh? (b) a dog weigh?

47. A boy is given a choice of six dozen dozen chocolate bars or a half a dozen dozen. Which is greater?

48. Two automobiles start at the same time, 280 miles apart, and travel toward each other. One travels at the rate of 40 miles per hour and the other at a rate of 30 miles per hour. How long will it be before they meet?

4.6. Short Methods of Division. Of the short methods of division, the only important one is the so-called Austrian method. This method, instead of our long method, is taught in almost all of the primary schools in Europe and Latin America. It eliminates the time required to write each partial product by performing the multiplication and subtraction mentally. This seems to imply that a great mental effort is required. The fact is that by using the balancing-additions method of subtraction (Section 2.5) it is quite easy and once mastered is, without doubt, much faster.

Example. Using the short method, divide 872 by 24.
Solution. The usual long method is shown at the left for purposes of comparison.

```
       36                      3                       36
   24) 872                 24) 872                 24) 872
       72                      15                      152
      ————                                              8
      152
      144
      ————
        8
```

The first digit in the quotient, obtained in the usual manner, is 3. Now $3 \times 4 = 12$; but, we cannot take 12 away from 7, so we add 10 and make it 17. Then $17 - 12 = 5$. Write 5 under the 7 in the dividend and carry 1, for we added 1 ten to 7. Now $3 \times 2 + 1$ (carried) $= 7$ and $8 - 7 = 1$. Write 1 under the 8 in the dividend. Thus, the first remainder is 15. Bring down the 2 as shown at the right, above. The next digit in the quotient is 6. Now $6 \times 4 = 24$. But we cannot take 24 away from 2, so we add 30 and make it 32. Then $32 - 24 = 8$. Write 8 under the 2 in 152 and carry 3 (since we added 3 tens to 2, getting 32). Now $6 \times 2 + 3$ (carried) $= 15$ and $15 - 15 = 0$. Thus, the remainder is 8.

This operation, like all other operations in arithmetic, should be performed, with practice, by immediately seeing the products and the remainders. Thus, considering the product of 24 by 3 and the succeeding subtraction, one should think 12, 17, 5. Write 5 under 7. Then think 6, 7, 8, 1. Write 1 under the 8 and bring down the 2. To consider the product of 24 by 6 and the corresponding subtraction, think 24, 32, 8. Write 8 under 2. Finally, think 12, 15, 15, 0.

Example. Using the short method, divide 1,678,218 by 82.
Solution. Here $16 \div 8 = 2$, so write 2 as the first digit in the quotient; $2 \times 2 = 4$, $7 - 4 = 3$; write 3 under the 7. Then $2 \times 8 = 16$ and $16 - 16 = 0$. Bring down the 8 from the dividend.

```
       20466
   82) 1,678,218
       38 2
        5 41
         498
           6
```

Since 82 does not go into 38, write 0 as the second digit in the quotient and bring down the 2. The next digit in the quotient is 4, because $38 \div 8$ is approximately 4. Then $4 \times 2 = 8$ which subtracted from 12 $(10 + 2)$ leaves 4. Write 4 under the 2 and carry 1. Now $4 \times 8 + 1$ (carried) $= 33$ and $38 - 33 = 5$ so write 5 under the 8 and bring down the 1. Since $54 \div 8$ is approximately 6, the next digit in the quotient is 6. Now $6 \times 2 = 12$ and 12 from 21 $(20 + 1)$ leaves 9, so write 9 under the 1 and carry 2. Then $6 \times 8 + 2$ (carried) $= 50$ and $54 - 50 = 4$. Write 4 under the 4 in 541 and bring down the 8. Finally, $49 \div 8$ is approximately 6, so write 6 as the last digit in the quotient. Then $6 \times 2 = 12$ which subtracted from 18 $(10 + 8)$ leaves 6. Write 6 under the 8 in 498 and carry 1. Now $6 \times 8 + 1$ (carried) $= 49$ and $49 - 49 = 0$. Thus, the remainder is 6.

Again, one should think as follows. Consider the product 82×2 and the succeeding subtraction. One should immediately see 4, 7, 3. Write 3 under the 7. Then think 16, 16, 0. Bring down the 8. The next digit in the quotient is 0. Bring down the 2 and consider the product 82×4 and the corresponding subtraction. Think 8, 12, 4. Write 4 under the 2 in 382. Then think 32, 33, 38, 5. Write 5 under the 8 in 382 and bring down the 1. Now consider the product 82×6 and the corresponding subtraction and think 12, 21, 9. Write 9 under the 1 in 541 and continue 48, 50, 54, 4. Write 4 under the 4 in 541 and bring down the 8. Finally, consider the product 82×6 and the corresponding subtraction to obtain 12, 18, 6. Write 6 under the 8 in 498. Then think 48, 49, 49, 0.

EXERCISE 15

Divide, using the short method.

1. $504 \div 6$	**6.** $3{,}185 \div 5$	**11.** $23{,}621 \div 58$
2. $329 \div 7$	**7.** $2{,}848 \div 32$	**12.** $38{,}860 \div 77$
3. $472 \div 8$	**8.** $3{,}116 \div 41$	**13.** $52{,}100 \div 123$
4. $5{,}061 \div 7$	**9.** $3{,}078 \div 54$	**14.** $324{,}200 \div 534$
5. $2{,}116 \div 4$	**10.** $3{,}936 \div 83$	**15.** $343{,}100 \div 682$

16. Find a number which added to its double gives 114.

17. (a) How long will it take to drive from New Orleans, La. to Dallas, Tex., a distance of 504 miles, if the driver averages 36 miles per hour? (b) If the automobile can travel 14 miles on a gallon of gasoline, how many gallons must be consumed for the trip?

18. A number is multiplied by 7 and 50 is added to the product. Then 103 is subtracted from this product and the difference is divided by 25, obtaining a quotient of 118. Find the number.

19. If it takes 12 one-cent stamps to make a dozen, how many two-cent stamps will it take?

20. There are fifteen ears of corn in a barn. How long will it take a squirrel to carry them all out if he carries out 3 ears a day?

4.7. Early Forms of Division. Probably the oldest method of division which used the Hindu-Arabic numerals can be attributed to Gerbert, who later became Pope Sylvester II (999).[#] This method is illustrated in the following example.

[#]Smith, *op. cit.*

Example. Using Gerbert's method, divide 792 by 18.
Solution. Express 18 as 20 − 2.

$$18 = 20 - 2\overline{)700 + 90 + 2}\big(30 + 10 + 3 + 1 = 44$$

$$\underline{600 - 60}$$
$$100 + 150 + 2 = 200 + 50 + 2$$
$$\underline{200 - 20}$$
$$70 + 2$$
$$\underline{60 - 6}$$
$$10 + 8 = 18 = 20 - 2$$
$$\underline{20 - 2}$$
$$0$$

Write the dividend and the divisor as shown above. Then since 20 is contained in 700 thirty times, write 30 as the first number in the quotient. Multiply 20 − 2 by 30 and write the product, 600 − 60, below 700 + 90. In subtracting 600 − 60, we are actually subtracting 540. Hence, if we take away 600, we must add 60. Thus if we subtract 600 − 60 from 700 + 90, we obtain a remainder of 100 + 150. Bring down the 2 and combine similar units to get 200 + 50 + 2. Now 200 ÷ 20 = 10, so write 10 as the second number in the quotient. Multiply (20 − 2) × 10 and write the product, 200 − 20, under 200 + 50 + 2. Subtract, obtaining 70 + 2. Since 20 goes into 70 three times, write 3 as the next number in the quotient. Multiply (20 − 2) × 3 and write the product, 60 − 6, below 70 + 2. Subtract, obtaining 10 + 8 = 18 = 20 − 2. Finally, 20 − 2 goes into 20 − 2 once with no remainder. Write 1 as the last number in the quotient. The quotient is 30 + 10 + 3 + 1 = 44.

The Galley or Scratch method was the most common form of division used before 1600. The name of the method is derived from the Italian word *galea* meaning a ship having three or four decks. The Treviso arithmetic (1478) gives the following example.

Example. Using the Galley method, divide 65,284 by 594.
Solution. Beginning at (1), write the divisor under the dividend so that

    (1)   (2)   (3)

the first digit in the divisor is below the first digit in the dividend. Since 594 goes into 652 once, write 1 to the right of 65284 as the first digit in the quotient. As shown in (2), multiply the first digit in the divisor by 1 and subtract the product, 5, from 6. Write the difference

1 over the 6 in the dividend and scratch this 6 and the 5 below it. Now find the difference $5 - (9 \times 1)$. Since 9 cannot be taken away from 5, borrow 1 unit from the 1 over the scratched 6, leaving 0, and add the 1 unit borrowed to 5, obtaining $10 + 5 = 15$. Subtract $15 - 9 = 6$. Write 6 over the 5 in the dividend and scratch the 1 over the 6 in the dividend, the 5 in the dividend, and the 9 in the divisor. Now, as shown in (3), subtract 4×1 from 2. But 4 cannot be subtracted from 2, so borrow 1 unit from 6, leaving 5, and add the 1 unit to 2 to obtain $10 + 2 = 12$. Scratch the 6 and write 5 over it. Then $12 - 4 = 8$. Write 8 over the 2 in the dividend and scratch the 2 and the 4 below it. Rewrite the divisor 594 one place to the right, as shown in (4). Now 594

```
        (4)                    (5)                     (6)

                                1                       15
         5                     53                      533
        168                   168                     16878
       63284  10             63284  109              63284  109
        594                   594                     594
         594                   594                     594
                               594                     594
```

does not go into 588; hence, write 0 as the next digit in the quotient and scratch 5, 9, and 4 in the divisor. Rewrite the divisor one place to the right, as shown in (5). Consider how many times 594 goes into 5884. The answer is 9, so write 9 as the next digit in the quotient. Multiply $9 \times 5 = 45$ and subtract this product from 8. Since this is impossible, we must borrow 4 units from the 5 above the 6, leaving 1. Scratch the 5, write 1 above it, and add the 4 units borrowed to 8, obtaining $40 + 8 = 48$. Now subtract $9 \times 5 = 45$ from 48 to get the difference, 3. Write 3 over the 8 in 168 and scratch this 8 and the 5 in 594. As shown in (6), multiply $9 \times 9 = 81$. Since 81 cannot be subtracted from 8, we must borrow 8 units to obtain a number larger than 81. First, borrow the 1 unit over the 5 in 53 to get $10 + 3 = 13$ units, and scratch this 1. Now borrow 8 units from the 13 units just obtained, leaving 5. Write 5 over the 3 in 53 and scratch this 3. Finally, adding the 8 units borrowed to 8, we obtain $80 + 8 = 88$ and the difference, 7 ($88 - (9 \times 9) = 88 - 81 = 7$), is written over the 8 in the dividend. Scratch the 8 in the dividend and the 9 in the divisor. Now multiply $9 \times 4 = 36$. But 36 cannot be taken away from 4, so to obtain a number larger than 36, borrow 4 units from the 7, leaving 3. Scratch the 7 and write 3 above it. Then add the 4 units borrowed to 4 to get $40 + 4 = 44$ and find the difference $44 - (9 \times 4) = 44 - 36 = 8$. Scratch the 4 in the dividend, write 8 above it, and scratch the 4 in the divisor. Thus, the quotient is 109 and the remainder is 538.

4.8. Checking Division. Since division is the inverse operation of multiplication, and since dividend = divisor × quotient + remainder, the obvious and usual check for division is to find the product of the divisor times the quotient, and add the remainder.

Example. Using the short method, divide 54,347 by 71, and check your answer by multiplication.

Solution.

$$
\begin{array}{r}
765 \\
71\overline{)54{,}347} \\
4\ 64 \\
387 \\
32
\end{array}
\qquad
\text{CHECK.}
\qquad
\begin{array}{r}
765 \\
71 \\
\hline
765 \\
53\ 55 \\
\hline
54\ 315 \\
32 \\
\hline
54{,}347
\end{array}
$$

The easiest and fastest way to check division is by casting out 9's or 11's. Keeping the following in mind—(1) the sum of the check numbers of the addends is equal to the check number of the sum,[||] (2) the product of the check numbers of the factors is equal to the check number of the product,[||] and (3) the fundamental relation of division stated above—it follows that the check number of the dividend is equal to the check number of the divisor times the check number of the quotient plus the check number of the remainder.

Example. Check the division performed in the preceding example by (a) casting out 9's; (b) casting out 11's.
Solution. From the preceding example,

$$54{,}347 = 71 \times 765 + 32.$$

(a) The check number of 54,347 is $5 + 4 + 3 + 4 + 7 = 23 = 2 + 3 = 5$; for 71, it is $7 + 1 = 8$; for 765, it is $7 + 6 + 5 = 18 = 1 + 8 = 9$ or 0; and for 32, it is $3 + 2 = 5$. Since the check number of 71 × the check number of 765 + the check number of 32 = $8 \times 0 + 5 = 5$, and the check number of 54,347 is also 5, the division is correct.

(b) The check number of 54,347 is $(7 + 3 + 5) - (4 + 4) = 15 - 8 = 7$; for 71, it is $1 - 7 = (1 + 11) - 7 = 12 - 7 = 5$; for 765, it is $(5 + 7) - 6 = 12 - 6 = 6$; and for 32, it is $2 - 3 = (2 + 11) - 3 = 13 - 3 = 10$. Since the check number of 71 × the check number of 765 + the check number of 32 is equal to

$$5 \times 6 + 10 = 30 + 10 = 40 \text{ or } 0 - 4 = (0 + 11) - 4 = 7$$

and the check number of 54,347 is also 7, the division is correct.

[||] For proof of this statement, see page 159.

EXERCISE 16

Divide, using Gerbert's or the Galley method. Check your answer by multiplying, casting out 9's, or casting out 11's.

1. 988 ÷ 19	7. 2,789 ÷ 38	13. 896 ÷ 14
2. 868 ÷ 28	8. 7,200 ÷ 17	14. 884 ÷ 34
3. 897 ÷ 39	9. 22,979 ÷ 98	15. 2,964 ÷ 52
4. 954 ÷ 18	10. 52,100 ÷ 99	16. 6,360 ÷ 76
5. 783 ÷ 29	11. 945 ÷ 45	17. 3,182 ÷ 67
6. 2,993 ÷ 48	12. 782 ÷ 23	18. 127,394 ÷ 253

19. A college bookstore bought 120 books at $4 each. Twelve of these books were sold for $3 each due to damage in handling. At what price should each of the remaining books be sold to make a total profit of $312?

20. A rancher bought 5 steers and 4 horses for $410. He then bought, at the same price, 15 steers and 6 horses for $990. How much did he pay for: (a) a horse? (b) a steer?

21. A passenger train leaves New York City for Harrisburg, Pa. and travels at the rate of 57 miles per hour. At the same time another train leaves Harrisburg, Pa. for New York City and travels at the rate of 32 miles per hour. If the distance between these two cities is 178 miles, how long will it be before the two trains meet?

22. Mr. Rogers has a balance of $3,578. If he buys merchandise worth $330 and with the remainder buys flour at $8 per barrel, how many barrels of flour can he buy?

23. Show that if twelve is divided into two equal parts, one of the two equal parts is equal to seven.

24. Show that if nine is divided into two equal parts, one of the two equal parts is equal to four.

25. There are six apples in a box. Divide these apples among six boys in such a way as to leave an apple in the box without cutting the apples.

Factoring

5.1. Factors. A factor (as defined on page 9) is any one of two or more numbers which are multiplied together to form a product. Taking 4 × 5 = 20 as an example, 4 is a factor of 20, and 5 is a factor of 20. It immediately follows, from the definition of division, that *every factor of a product is an exact divisor of that product.* Hence, the factors of a given number are all the numbers which can be divided into that given number with a remainder of zero. Thus, the factors of 20 are 1, 2, 4, 5, 10, and 20. Since any given number is exactly divisible by 1 and by itself, these factors, being universally known, are not usually stated. Following custom then, we say the factors of 20 are 2, 4, 5, and 10, leaving it understood that 1 and 20 are also factors.

5.2. Classification of Integers. Any integer greater than one that does not have any factors except 1 and itself is called a *prime number.* The number 1 is excluded because 1 × 1 × 1, and so on, equals 1. Thus, the prime numbers from 1 to 25 are 2, 3, 5, 7, 11, 13, 17, 19, and 23.

The factors of a number which are prime numbers are called the *prime factors* of that number. For example, 2, 3, 4, and 6 are factors of 12, for 12 = 2 × 6 and 12 = 3 × 4. Of these factors, only 2 and 3 are prime numbers; hence, 2 and 3 are the prime factors of 12. Moreover, 12 expressed in terms of its prime factors is 12 = 2 × 2 × 3.

Any integer, except 0 and 1, that can be obtained by the multiplication of other integers is a *composite number.* In other words, a composite number is any integer, except 0 and 1, that is not prime. The numbers 0 and 1 are excluded because, as already stated in Section 3.3, 0 × N = 0 no matter what the value of N may be, and the product of 1 by any number is the number itself.

Any integer that has 2 as a factor is an *even number*. Integers not divisible by 2 are called *odd numbers*.

From these definitions, we can immediately deduce that all prime numbers, except 2, are odd, for by definition, all even numbers contain 2 as a factor. However, not all the odd numbers are prime. For example, 21 is odd but composite, for $21 = 3 \times 7$.

As will be shown in Section 5.4, any number whose units digit is even, is divisible by 2, and any number whose units digit is 5, is divisible by 5. Hence, all prime numbers, except 2 and 5, must have 1, 3, 7, or 9 as their units digit.

5.3. Factoring. The process of finding the factors of a composite number is called *factoring*. Note that any exact divisor of the number to be factored is one factor, and the quotient is another factor. For if $a \div b = c$, then, since division is the inverse operation of multiplication, $a = b \times c$, and b and c are factors of a. For example, $21 \div 7 = 3$; hence, $21 = 3 \times 7$, and 3 and 7 are factors of 21. Thus we can obtain the factors of a number by dividing successively by its exact divisors. These exact divisors or factors can be found by using a trial and error method until the factors are determined. But since this can cause unnecessary work, it is desirable to develop practical rules of divisibility.

5.4. Divisibility of Numbers. Any number can be written as the sum of its units digit and the product of all its other digits by 10. For example, 2,348 can be written $(234 \times 10) + 8$. Hence, if u represents the units digit of a number N, and D denotes all its other digits, then

$$N = (D \times 10) + u.$$

Moreover, if N is divided by any integer b

$$N \div b = [(D \times 10) + u] \div b$$

then, by law 5, page 60,

$$N \div b = [(D \times 10) \div b] + (u \div b).$$

It follows that the number N is divisible by b if $(D \times 10)$ and u are each divisible by b. No matter what the value of D may be, the product $(D \times 10)$ is divisible by 2 and also by 5, for these numbers are exact divisors of 10. Therefore, if u (the units digit) is divisible by 2 or 5, the number is divisible by 2 or 5.

1) *A number is divisible by 2 if its units digit is divisible by 2.*

2) *A number is divisible by 5 if its units digit is divisible by 5, that is, if its units digit is 5.*

Furthermore, if the units digit is 0, and the other digits of the number are represented by *D*, then a number *N* can be written as $D \times 10$. For example, $2,340 = 234 \times 10$. It immediately follows that since 2 and 5 are exact divisors of 10, no matter what the value of *D* may be, the number is divisible by 2, by 5, and of course by 10. Hence,

3) *If the units digit of a number is 0, the number is divisible by 2, by 5, and by 10.*

4) *A number is divisible by 4 if the two right-hand digits are zeros, or if these digits form a number which is divisible by 4.* If the two right-hand digits are zeros, the number is equal to a number of hundreds. And since 100 is divisible by 4, any number of hundreds is divisible by 4. Moreover, if the two right-hand digits form a number which is divisible by 4, the number can be written as a certain number of hundreds plus the number formed by the two right-hand digits. Since both of these are divisible by 4, by law 5, page 60, the number is divisible by 4.

5) *A number is divisible by 8 if the three right-hand digits are zeros, or if these digits form a number which is divisible by 8.* If the three right-hand digits are zeros, the number is equal to a number of thousands. And since 1,000 is divisible by 8, any number of thousands is divisible by 8. Furthermore, if the three right-hand digits form a number which is divisible by 8, the number can be written as a certain number of thousands plus the number formed by the three right-hand digits. Since both of these numbers are divisible by 8, by law 5, page 60, the number is divisible by 8.

6) *A number is divisible by 9 if the sum of its digits is divisible by 9.** Consider the number $4,635 = 4,000 + 600 + 30 + 5$. Each one of these parts can be written as shown below.

$$4,000 = 4 \times 1,000 = 4 \times (999 + 1) = 4 \times 999 + 4$$
$$600 = 6 \times 100 \quad = 6 \times (99 + 1) \quad = 6 \times 99 \quad + 6$$
$$30 = 3 \times 10 \quad = 3 \times (9 + 1) \quad = 3 \times 9 \quad + 3$$
$$5 \qquad\qquad\qquad\qquad\qquad\qquad\qquad = \qquad 5$$

Hence we can write

$$4,635 = (4 \times 999) + (6 \times 99) + (3 \times 9) + (4 + 6 + 3 + 5).$$

*An algebraic proof of this rule is given on page 159.

Now each of the first three terms of this expression, namely (4 × 999), (6 × 99), and (3 × 9), are evidently divisible by 9. Hence, if the number 4,635 is divided by 9, the remainder, if any, will be due to the division of the sum 4 + 6 + 3 + 5 by 9. It is easily seen that all powers of 10 less 1 are divisible by 9. Thus, $10 - 1 = 9$, $10^2 - 1 = 99$, $10^3 - 1 = 999$, and so on, are all divisible by 9. Moreover, since any number in our system can be expressed in terms of powers of 10 less 1, as shown above, then *any number divided by 9 leaves the same remainder as the sum of its digits divided by 9.* Therefore, if the sum of the digits of a number is divisible by 9, then by law 5, page 60, the number is divisible by 9. Now in the number 4,635 the digits 4 + 6 + 3 + 5 = 18 and 18 ÷ 9 = 2; therefore, 4,635 is divisible by 9.

7) *A number is divisible by 3 if the sum of its digits is divisible by 3.* This rule follows from the preceding demonstration where it was shown that any number can be expressed as the sum of a number of terms all of which are divisible by 9 plus the sum of its digits. Since 3 is a factor of 9, all the terms that are divisible by 9 are divisible by 3. Therefore, by law 5, page 60, a number is divisible by 3 if the sum of its digits is divisible by 3.

8) *A number is divisible by 6 if it is an even number and if the sum of its digits is divisible by 3.* Recall that if a number is even, then (by rule 1) it is divisible by 2, and if the sum of its digits is divisible by 3, then (by rule 7) it is divisible by 3. It immediately follows that since the number is divisible by both 2 and 3, it is divisible by their product 2 × 3 = 6.

9) *A number is divisible by 11 if the difference between the sum of its digits in the odd places and the sum of its digits in the even places is divisible by 11 or is equal to 0.*† The demonstration of this rule is based on the fact that all odd powers of 10 plus 1 and all even powers of 10 less 1 are divisible by 11. Thus, $10 + 1 = 11$, $10^3 + 1 = 1,001$, $10^5 + 1 = 100,001$, and so on, are all divisible by 11 as are the following differences, $10^2 - 1 = 99$, $10^4 - 1 = 9,999$, and so on. Now consider the number 83,765 = 80,000 + 3,000 + 700 + 60 + 5 and rewrite each of its parts as follows.

$$
\begin{array}{llll}
5 = & & & + 5 \\
60 = 6 \times 10 & = 6 \times (11 - 1) & = 6 \times 11 & - 6 \\
700 = 7 \times 100 & = 7 \times (99 + 1) & = 7 \times 99 & + 7
\end{array}
$$

†An algebraic proof of this rule is given on page 160.

$$3,000 = 3 \times 1,000 = 3 \times (1,001 - 1) = 3 \times 1,001 - 3$$
$$80,000 = 8 \times 10,000 = 8 \times (9,999 + 1) = 8 \times 9,999 + 8$$

Consequently we can write

$$83,765 = (8 \times 9,999) + (3 \times 1,001) + (7 \times 99)$$
$$+ (6 \times 11) + (8 + 7 + 5) - (3 + 6).$$

Now the first four terms of this expression, namely $(8 \times 9,999)$, $(3 \times 1,001)$, (7×99), and (6×11), are each divisible by 11. Hence, if the number is divided by 11, the remainder, if any, will be due to the division of the difference between the sum of the digits in the odd places and those in the even places by 11. Moreover, since any number in our system can be expressed in powers of 10, as shown above, then *any number divided by eleven leaves the same remainder as the difference between the sum of its digits in the odd places and those in the even places divided by 11.* Accordingly, if $(8 + 7 + 5) - (3 + 6) = 20 - 9 = 11$ is divisible by 11, then by law 5, page 60, the number 83,765 is divisible by 11. Since 11 is an exact divisor of itself, 83,765 is divisible by 11. This rule simply determines the remainder after casting out the elevens in the number, so the addition of 11 to the number, or any of its multiples, will not change the remainder. Therefore, if the sum of the digits in the even places is greater than the sum of the digits in the odd places, add 11, or any multiple thereof, to the sum of the digits in the odd places to obtain the correct remainder.

Example. Express 468 in terms of its prime factors.
Solution. The two right-hand digits of 468 form the number 68 which is divisible by 4, for $68 \div 4 = 17$. Hence, by rule 4, the number 468 is divisible by 4. Thus, $468 \div 4 = 117$ and $468 = 4 \times 117$.

The sum of the digits of 117, that is, $1 + 1 + 7 = 9$, is divisible by 9, for $9 \div 9 = 1$. Hence, by rule 6, the number 117 is divisible by 9. Thus, $117 \div 9 = 13$ and $117 = 9 \times 13$.

Since 13 is a prime number, there are no more factors. Hence, $468 = 4 \times 117 = 4 \times 9 \times 13$. But $4 = 2 \times 2$ and $9 = 3 \times 3$; therefore,

$$468 = 2 \times 2 \times 3 \times 3 \times 13. \quad Ans.$$

Example. Express 3,315 in terms of its prime factors.
Solution. The units digit of the number is 5; hence, by rule 2, the number is divisible by 5. Thus, $3,315 \div 5 = 663$ and $3,315 = 5 \times 663$.

The sum of the digits of 663 is $6 + 6 + 3 = 15$ and 15 is divisible by 3, for $15 \div 3 = 5$. Hence, by rule 7, the number 663 is divisible by 3. Thus, $663 \div 3 = 221$ and $663 = 3 \times 221$.

None of the eight rules given apply to 221, so try the prime numbers; see problem 1 of Exercise 17. Thus, dividing by 13, we obtain $221 \div 13 = 17$ and $221 = 13 \times 17$.

Since these numbers are prime numbers, there are no more factors. Therefore,

$$3,315 = 5 \times 663 = 5 \times 3 \times 221 = 5 \times 3 \times 13 \times 17. \quad Ans.$$

EXERCISE 17

1. Find all the prime numbers from 1 to 100. *Hint:* Write the numbers from 1 to 100 in rows of 10 each. Cancel all the even numbers except 2, every third odd number after 3, every fifth odd number after 5, every seventh odd number after 7, and so on. This method is known as the *sieve of Eratosthenes* (230 B.C.).

Express each of the following numbers in terms of its prime factors.

2. 42	**4.** 87	**6.** 192	**8.** 403	**10.** 770	**12.** 2,002	**14.** 1,512
3. 75	**5.** 94	**7.** 185	**9.** 201	**11.** 1,755	**13.** 1,260	**15.** 16,830

5.5. Greatest Common Divisor. The word *divisor* as used in the expression, *greatest common divisor,* means an exact divisor. As mentioned in Section 5.1, an exact divisor of a number is a factor of that number. Hence, a *common divisor* of two or more numbers is a factor common to all of them. For example, 420 and 108 have the numbers 2, 4, 3, 6, and 12 as common divisors. The *greatest common divisor* of two or more numbers is the greatest number which is a factor common to all of them. Thus, 420 and 108 have the numbers 2, 3, 4, 6, and 12 as common divisors but 12 is the greatest common divisor (G. C. D.) because 12 is the greatest number which is a factor of both 420 and 108. We can note this by writing: G. C. D. (420 & 108) = 12.

To find the greatest common divisor of two or more numbers, first express the given numbers in terms of their prime factors. Then, choose all the prime factors common to all the numbers the *least number of times* they appear as factors of any one number. Since these prime factors are all common factors of the numbers, their product is the greatest common divisor of the numbers.

Example. Find the G. C. D. of 420 and 108.
Solution. $420 = 2 \times 2 \times 3 \times 5 \times 7.$
$108 = 2 \times 2 \times 3 \times 3 \times 3.$

The prime factors common to both numbers are 2 and 3. The least number of times 2 appears as a factor of any one number is twice.

(Twice, it is a factor of both 420 and 108.) The least number of times 3 appears as a factor of any one number is once as a factor of 420. Therefore,

G. C. D. (420 & 108) = 2 × 2 × 3 = 12. *Ans.*

The greatest common divisor can also be found without finding the prime factors of each number, by the method illustrated in the following example.

Example. Find the G. C. D. of 60, 420, and 660.

Solution. Write the numbers in a row as shown at the left. Since all numbers contain 10 as a factor, divide by 10 and write the remainders 6, 42, and 66 under the corresponding numbers. These remainders contain 6 as a factor. Hence, divide by 6 obtaining the remainders 1, 7, and 11. Since these remainders have no common factor, the

$$
\begin{array}{r|rrr}
10 & 60 & 420 & 660 \\
\hline
6 & 6 & 42 & 66 \\
\hline
& 1 & 7 & 11
\end{array}
$$

G. C. D. (60, 420, & 660) = 10 × 6 = 60. *Ans.*

When the factors of the numbers are not easy to obtain, the greatest common divisor can be found by the Euclidean Algorithm,‡ named after Euclid (300 B.C.). This method consists in successively dividing the greater number by the smaller, then the smaller number by the remainder of the first operation, then the remainder of the first operation by the remainder of the second, and so on, until a remainder of zero is obtained. The last divisor is the greatest common divisor.

Example. Using the Euclidean Algorithm, find the G. C. D. of 252 and 132.

Solution. Divide 252 by 132 obtaining a quotient of 1 and a remainder of 120, as shown at the left. Then divide 132 by 120 obtaining a quotient of 1 and a remainder of 12. Finally, divide 120 by 12 obtaining a quotient of 10 and a remainder of 0. The last divisor is 12. Therefore,

$$
\begin{array}{r}
1 \\
132\overline{)252} \\
\underline{132}\quad 1 \\
120\overline{)132} \\
\underline{120}\quad 10 \\
12\overline{)120} \\
120
\end{array}
$$

G. C. D. (252 & 132) = 12. *Ans.*

Example. Three blocks of cheese weigh 650 lbs., 680 lbs., and 760 lbs., respectively. For the purpose of packing and to avoid waste, the cheese is to be cut into pieces of equal weight. The weight of these equal pieces is to be as great as possible. How much should each piece weigh?

‡ For a proof of the Euclidean Algorithm, see page 161.

Solution.   For the pieces to be of equal weight and for the weight to be
as great as possible, the G. C. D. of 650, 680,

$$
\begin{array}{r|rrr}
10 & 650 & 680 & 760 \\
\hline
 & 65 & 68 & 76 \\
\end{array}
$$

and 760 must be found. All these numbers have
10 as a common factor. Dividing each of these
numbers by 10, we obtain remainders of 65, 68,
and 76. Now 65 = 5 × 13, but since these numbers are not factors of
68 or of 76, the G. C. D. (650, 680, & 760) = 10 lbs.   *Ans.*

EXERCISE 18

Find the G. C. D.

1. 27 & 36	**5.** 162 & 216	**9.** 16, 64, & 96	**13.** 24, 80, & 160
2. 54 & 72	**6.** 72 & 222	**10.** 90, 180, & 945	**14.** 125, 550, & 1,050
3. 152 & 184	**7.** 324 & 342	**11.** 72, 216, & 128	**15.** 36, 72, 108, & 144
4. 525 & 675	**8.** 232 & 104	**12.** 60, 320, & 420	**16.** 18, 54, 108, & 162

17. Two ribbons, 48 and 36 yards long, are to be cut into pieces of equal
length. The length of these equal pieces is to be as great as possible,
with no waste. What should the length of each piece be?

18. What is the size of the longest unmarked measure that can be used
to measure exactly 48, 138, 582, and 786 yards of material?

19. A paymaster has a certain number of 1, 2, 5, 10, and 20 dollar bills.
He has many more bills of one denomination than of the others and
wants to pay only with this particular kind of bill if possible. He
makes three payments of $210, $715, and $425, respectively, using
only bills of one denomination. What did he use?

20.   A girl with lovely eyes
went out to view the skies.
She saw an apple tree with apples on it.
She neither took apples nor left apples.
How many apples were on the tree?

5.6. Least Common Multiple.   A number which is the product of
a given number and another factor is a *multiple* of the given
number. For example, 12 is a multiple of 2, 3, 4, and 6 and also
of 1 and 12 but, since every number is a multiple of 1 and itself,
these numbers are not usually stated as multiples.   From the
definition of multiple, it follows that a multiple of a given number
contains the given number as a factor and hence, the multiple is
exactly divisible by that given number. A *common multiple* of two
or more given numbers is any number which contains all the
given numbers as factors. For example, 36, 24, and 12 are com-

mon multiples of 2, 3, 4, and 6 for each of the numbers—36, 24, and 12—contains all of the numbers 2, 3, 4, and 6 as factors. The *least common multiple* (L. C. M.) of two or more given numbers is the *smallest* number which contains all the given numbers as factors. Thus, although 36, 24, and 12 are all common multiples of 2, 3, 4, and 6, the least common multiple of these numbers is 12 because it is the smallest number which contains all of the numbers 2, 3, 4, and 6 as factors. We can note this fact by writing: L. C. M. (2, 3, 4, & 6) = 12.

To find the least common multiple of two or more given numbers, first express each of the given numbers in terms of their prime factors. Then choose *all* the prime factors the *greatest* number of times they appear as factors of any one number. Since these prime factors are all factors of the given numbers, their product is the smallest number which contains all the given numbers as factors, that is, the least common multiple.

Example. Find the L.C.M. of 80 and 300.
Solution.   80 = 2 × 2 × 2 × 2 × 5.
                300 = 2 × 2 × 3 × 5 × 5.

All the numbers which appear as prime factors are 2, 3, and 5. The greatest number of times 2 appears as a factor of any one number is 4, as a factor of 80. The greatest number of times 3 appears as a factor is 1, as a factor of 300. Finally, the greatest number of times 5 appears as a factor is 2, as a factor of 300. Therefore,

  L. C. M. (80 & 300) = 2 × 2 × 2 × 2 × 3 × 5 × 5 = 1,200. *Ans.*

The least common multiple can also be found without finding the prime factors of each number by the following method.

Example. Find the L.C.M. of 8, 12, 30, and 48.
Solution. Write the numbers in a row as shown at the left. Whenever two or more of the numbers have a common factor, divide by this factor and carry down, unchanged, any number not exactly divisible by it. Thus, 12, 30, and 48 have 3 as a factor. Divide by 3 and write the remainders, 4, 10, and 16, under the corresponding numbers, carrying down the 8. Now 8, 4, and 16 have 4 as a common factor. Divide by 4 and write the remainders, 2, 1, and 4, under the corresponding numbers, carrying down the 10. Finally, 2, 10, and 4 have 2 as a common factor. Divide by 2 and write the remainders, 1, 5, and 2, under the correspond-

```
3 | 8  12  30  48
4 | 8   4  10  16
2 | 2   1  10   4
    1   1   5   2
```

ing numbers, carrying down the 1. Since none of the remainders have a common factor, the L. C. M. (8, 12, 30, & 48) equals

$$3 \times 4 \times 2 \times 1 \times 1 \times 5 \times 2 = 240. \quad Ans.$$

If the factors of the numbers are not easily obtainable, the least common multiple can be found by using the Euclidean Algorithm to determine the greatest common divisor, and then reasoning as follows. The least common multiple of two given numbers is, by definition, composed of all the factors of the smaller of the two given numbers and all the factors of the larger given number which are not contained in the smaller. The greatest common divisor contains all the prime factors contained in both of the given numbers the least number of times they appear as factors of any one number. Hence, if the larger of the two given numbers is divided by the greatest common divisor, the quotient will consist of the factors of the larger number not contained in the smaller. Thus if we multiply the smaller number by the quotient just obtained, we get a number that contains all the prime factors of the two given numbers the greatest number of times they appear as factors of any one number, that is, the least common multiple. Therefore, the least common multiple of two given numbers is equal to the product of the smaller given number by the quotient obtained by dividing the larger given number by their greatest common divisor. Hence, to find the least common multiple of two given numbers, divide their product by their greatest common divisor.

If there are more than two given numbers, find the least common multiple of any two of the given numbers, then find the L. C. M. of the least common multiple just obtained and another of the given numbers, and so on.

Example. Using the G. C. D. find the L. C. M. of 30, 260, and 312.
Solution. Using the Euclidean Algorithm find, as shown at the left, the greatest common divisor of 30 and 260. Thus,

$$G. C. D. (30 \& 260) = 10$$

The L. C. M. of 30 and 260 must be composed of all the factors of 30 and 260 not contained in 30. If we divide 260 by the G. C. D., we shall obtain the factors of 260 not belonging to 30. Therefore,

$$L. C. M. (30 \& 260) = 30 \times \frac{260}{10} = 780$$

Using 780 as one number and 312 as the other, we find, as shown at the left, that

$$\begin{array}{r} 2 \\ 312\overline{)780} \\ \underline{624} \quad 2 \\ 156\overline{)312} \\ \underline{312} \end{array}$$

G. C. D. (312 & 780) = 156

Hence,

L. C. M. (312 & 780)

$$= 312 \times \frac{780}{156} = 2 \times 780 = 1,560$$

Therefore,

$$\text{L. C. M.}(30, 260, \& \ 312) = 1,560. \quad \textit{Ans.}$$

Example. Three ships leave New York for Le Havre, France, on the same day. It takes the first ship 12 days, the second ship, 16 days, and the third ship, 20 days to make a round trip. (a) How many days will elapse before they again leave New York on the same day? (b) How many trips will each have made in the interval?

Solution. (a) The number of days that will elapse before the ships leave New York on the same day is the smallest number that contains all of these numbers as factors or the L. C. M. From the computation shown at the left, the

$$\begin{array}{r|ccc} 4 & 12 & 16 & 20 \\ \hline & 3 & 4 & 5 \end{array}$$

$$\text{L. C. M. } (12, 16, \& \ 20) = 4 \times 3 \times 4 \times 5 = 240 \text{ days.} \quad \textit{Ans.}$$

(b) The first ship makes 240 ÷ 12 = 20 trips.

The second ship makes 240 ÷ 16 = 15 trips.

The third ship makes 240 ÷ 20 = 12 trips.

EXERCISE 19

Find the L. C. M.

1. 60 & 81	**5.** 1,755 & 4,875	**9.** 100, 525, 560, & 875
2. 24 & 36	**6.** 9, 12, 21, & 30	**10.** 45, 320, 945, & 986
3. 70 & 130	**7.** 34, 58, 290, & 493	**11.** 39, 65, 78, 90, & 104
4. 624 & 702	**8.** 66, 99, 132, & 267	**12.** 44, 121, 1,100, 100, & 88

13. Three salesmen make trips at regular intervals: the first, every 7 days; the second, every 14 days; the third, every 21 days. (a) If they leave the office on the same day, how many days will elapse before they again leave the office on the same day? (b) How many trips will each have made in that interval?

14. What is the shortest piece of rope that can be cut evenly into lengths of either 12, 18, or 30 feet?

15. Find the smallest number which divided by each of the integers, 2, 3, 4, 5, 6, 7, 8, 9, and 10, will give, in each case, a remainder which is 1 less than the divisor.

5.7. Involution. If all the factors of a number are equal, the product is called the *power* of that factor (page 10). Thus, since $4 \times 4 \times 4 = 64$, the product 64 is said to be the third power of 4. This notation, namely $4 \times 4 \times 4$, is clumsy and if used, would slow up computation considerably. Instead of writing $4 \times 4 \times 4 = 64$, the same operation can be expressed as $4^3 = 64$. The small 3, written above and slightly to the right of 4, is called an *exponent*. It indicates how many times 4, which is called the *base*, is to be taken as a factor. The product of 4^3, namely 64, is called the *power*. In general, if N, a, and m are any integers where $N = a^m$, then a is the base, m is the exponent, and N is the power.

It is evident that $4^1 = 4$. Thus, whenever a number appears without an exponent, that number, by implication, has an exponent of 1. Hence, any number $a = a^1$. Although power usually refers to the product obtained by multiplying a number by itself a certain number of times, the word power can also be used to refer to the exponent. Thus, 5^4 is read five raised to the fourth power.

Since the area of a square is the product of two equal quantities, namely the length and the width, the result of raising a number to the second power is usually called the *square* of that number. Thus, 4^2 is read four squared. Similarly, since the volume of a cube is the product of three equal quantities, that is the length, the width, and the height, the result of raising a number to the third power is usually called the *cube* of that number. For example, 4^3 is read four cubed.

Example. (a) Express in exponential form, seven raised to the fifth power. (b) Express 5^3 as a product of its factors and find the power.
Solution. (a) 7^5. (b) $5^3 = 5 \times 5 \times 5 = 125$.

In finding the square of a number it is sometimes convenient to express the number in terms of its units, tens, and so on. A number of two digits can be expressed as the sum of a number of tens, say a, and some units, b. Figure 20 shows a square whose sides are equal to $a + b$. It is easily seen, from this figure, that the area of the square whose side is $a + b$ is equal to the area of a square whose side is a, that is a^2, plus the sum of the areas of two rectangles whose sides are a and b, namely $2ab$, plus the area of a square whose side is b, that is b^2. Thus,

$$(a + b)^2 = a^2 + 2ab + b^2$$

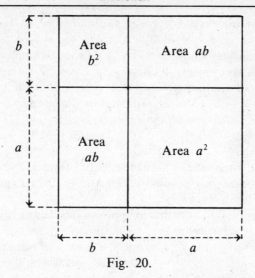

Fig. 20.

In general, the square of the sum of two numbers is equal to the square of the first, plus twice the product of the first and second, plus the square of the second. In other words, the square of a number of two digits is equal to the square of the tens, plus twice the product of the tens and the units, plus the square of the units.

Example. Find the square of 35.
Solution. $35 = 30 + 5$. Hence,

$$(35)^2 = (30 + 5)^2 = (30)^2 + 2 \times 30 \times 5 + (5)^2$$
$$= 900 + 300 + 25 = 1,225. \quad \textit{Ans.}$$

Our present notation for exponents began with Descartes (1637); its development arose from the study of equations. Hence, the rules of operation governing exponents are usually discussed in connection with the study of algebra.§ In scientific work, operations with very large numbers frequently occur. Instead of writing these numbers in full, the numbers are expressed in terms of powers of 10. For example, 27,800,000 is much easier to handle if written 278×10^5 for the latter notation avoids using long rows of zeros in writing and performing an operation.

§See Mira and Hartmann, *Business Mathematics* (Princeton: Van Nostrand, 1955).

By definition, $4^3 = 4 \times 4 \times 4$ and $4^2 = 4 \times 4$. Hence, $4^3 \times 4^2 = (4 \times 4 \times 4) \times (4 \times 4) = 4 \times 4 \times 4 \times 4 \times 4$. The last notation indicates 4 is to be taken as a factor five times. By definition of an exponent, 4 taken as a factor five times can be written 4^5. Therefore, $4^3 \times 4^2 = 4^5$. Note that the sum of the exponents $3 + 2 = 5$ is the exponent of the product. Similarly, where a, m, and n represent integers,

$$a^m \times a^n = a^{m+n}.$$

In general, when numbers in exponential form are expressed to the same base, their product is the sum of their exponents expressed to that same base.

It is important to note that in applying this or any other rule of exponents, the *number of the base must be the same*. If the meaning of an exponent is kept in mind, it is easy to see that $4^3 \times 5^2$ does not equal 4^5 or 5^5. Actually, $4^5 = 1,024$ and $5^5 = 3,125$, whereas $4^3 \times 5^2 = 4 \times 4 \times 4 \times 5 \times 5 = 1,600$.

Example. Find the product of 3^2 and 3^3.
Solution. $3^2 \times 3^3 = 3^5 = 3 \times 3 \times 3 \times 3 \times 3 = 9 \times 9 \times 3 = 81 \times 3 = 243$. *Ans.*

Example. Multiply 27,000,000 by 1,700,000.
Solution. $27,000,000 = 27 \times 10^6$ and $1,700,000 = 17 \times 10^5$. Hence,

$$27,000,000 \times 1,700,000 = (27 \times 10^6) \times (17 \times 10^5)$$
$$= 27 \times 17 \times 10^6 \times 10^5 = 459 \times 10^{11}$$
$$= 45,900,000,000,000. \quad Ans.$$

5.8. Evolution. Factoring was defined, in Section 5.3, as the process of finding the factors of a composite number. If all the numbers in our system are taken into account, any number can be expressed as the product of several equal factors. The process of finding one of the several equal factors of a number is called *evolution*. One of the several equal factors that will produce a number is called a *root* of that number. For example, since $4^2 = 4 \times 4 = 16$, 4 is a root of 16 for it is one of the two equal factors that will produce 16. The number of equal factors that produce a number determines the root to be found. Thus, one of the two equal factors that will produce a number is a *square root* of that number; a *cube root* is one of the three equal factors that will produce a number; a *fourth root*, one of the four equal factors; and so on. Thus, since $4 \times 4 = 16$, 4 is the square root of 16; the number 3 is the cube root of 27 because $3 \times 3 \times 3 = 27$. The symbol $\sqrt{}$, called a *radical sign*, is used to indicate a root. The number whose root is to be found is written under the radical sign and is called the *radicand*. The number indicating the root to be found is written above and to the left of the radical sign and is called the *index*. Hence, in the expression $\sqrt[3]{27}$, the number 27 is the radicand, and the number 3 is the index, and the expression is read "the cube root of 27."

Since a root of a number is *one* of the *several* equal factors that can produce that number, the least number of equal factors that can produce a number is two. Thus, the smallest index is 2. For this reason, and because the radical sign was probably used only for the square root before it was applied to higher roots, no index is needed to indicate the square root of a number. Thus, the square root of 16 is simply written $\sqrt{16}$.

The method for finding the square root of a number can be developed from the fact that the square of the sum of two numbers is equal to the square of the first plus twice the product of the first and second, plus the square of the second. That is, if a and b are any two numbers,

$$(a + b)^2 = a^2 + 2ab + b^2$$

Consider the problem of finding $\sqrt{576}$. The square root of this number will consist of two digits, for 576 is a number between $10^2 = 100$ and $100^2 = 10,000$. Hence, the square root of this number can be expressed as the sum of two numbers, namely a certain number of tens and a certain number of units. Therefore, if a denotes the number of tens and b represents the number of units, we can write

$$576 = (a + b)^2 = a^2 + 2ab + b^2$$

Now the greatest number of tens whose square is either equal to or less than 576 is 2 tens for $(20)^2 = 400$. Subtracting this square we obtain $576 - 400 = 176$. This number, 176, is equal to twice the product of the tens and the units plus the square of the units. That is, $176 = 2ab + b^2$. Since a represents the number of tens, and this is known to be 2, then $2a = 2 \times (2 \times 10) = 40$. But twice the product of the tens by the units is always greater than the square of the units, so that 176 primarily consists of twice the product of the tens by the units. Now since twice the product of the tens is 40, then $2ab = 40b$, so that 176 is approximately $40b$. Therefore, if we divide 176 by 40, we can find the units digit. Since 40 is contained in 176 four times, the units digit is 4 and $\sqrt{576} = 24$.

Any integer between 1 and 10 consists of *one* digit. The square of any of these integers will lie between $1^2 = 1$ and $10^2 = 100$ and hence, will consist of *one* or *two* digits. Any integer between 10 and 100 consists of *two* digits. The square of any of these integers will lie between $10^2 = 100$ and $100^2 = 10,000$ and hence, will consist of *three* or *four* digits, and so on. Therefore, the square of a number consists of twice as many digits as the number, or one less than twice the number of digits. It follows then that to find the square root of a number, groups of two digits each (beginning with the units digit) can be indicated, considering the single digit at the left, if any, as a group.

The results of the preceding discussion can now be used to develop a practical method of finding the square root of a number.

Example.   Find the square root of 77,284.
Solution.

1) Beginning with the units digit, separate the number into groups of two digits each, as shown in (a). The single digit 7 at the left is considered as a group.

```
        (a)                    (b)
         2                      2  9
   2 | 7'72'84           2 | 7'72'84
     | 4                   | 4
     |----                 |----
     | 3 72             49 | 3 72
                           | 4 41
```

2) Find the greatest number whose square is either equal to or less than 7. Since the answer is 2, write 2 above the radicand as the first digit of the root and write it again at the left of the radicand. Square 2, that is, get the product $2 \times 2 = 4$ and write it under the 7. Subtract $7 - 4 = 3$. Write 3 under the 4 and bring down the next group 72, obtaining 372.

3) As shown in (b), multiply the first digit in the root by 2. Thus, $2 \times 2 = 4$. Write 4 to the left of 372. Find how many times 4 is contained in 372 exclusive of the right-hand digit in 372. That is, find how many times 4 is contained in 37. The answer is 9. Write 9 as the second digit of the root and write it also to the right of 4, forming the number 49. Multiply 49 by 9 to get 441. Write this product, 441, under 372. Evidently, 441 is too large. Erase the 9 above the radicand, the 9 next to the

4, and also the 441. As shown in (c), try 7 as the second digit of the root. Write 7 next to the 2 in the root and also next to the 4, obtaining 47.

```
        (c)                      (d)
      2  7                     2  7  8
   2 | 7'72'84              2 | 7'72'84
     4                        4
   ―――――――                  ―――――――
  47 | 3 72              47 | 3 72
       3 29                   3 29
     ―――――――               ――――――――
         43 84           548 | 43 84
                               43 84
```

Find the product $47 \times 7 = 329$. Write this product under 372 and find the difference $372 - 329 = 43$. Bring down the next group, 84, to form the number 4384.

4) As shown in (d), multiply all the digits in the root, that is, 27, by 2. Write the product $27 \times 2 = 54$ at the left of 4384. Find how many times 54 is contained in 4384 exclusive of the right-hand digit 4. That is, find how many times 54 is contained in 438. The answer is 8. Write 8 as the next digit of the root and also write it next to 54 forming 548. Multiply $548 \times 8 = 4384$. Write this product under 4384 and subtract, obtaining a remainder of 0. Therefore, $\sqrt{77,284} = 278$. *Ans.*

Note: If the trial divisor is not contained in the dividend, place a zero in the root, bring down the next group, and double the root already found for the new trial divisor.

EXERCISE 20

Compute the value of each of the following.

1. 2×3^2
2. $(2 \times 3)^2$
3. $2^3 \times 2^4$
4. $3^2 \times 5^3 \times 3$
5. $7^2 \times 6^2$
6. $(7 \times 6)^2$
7. 7×6^2
8. $4^2 \times 5^3$
9. $4^2 \times 4^3$
10. $5^2 \times 5^3$
11. $3,500,000 \times 170,000$
12. $41,000,000 \times 23,000$
13. $8,100,000 \times 9,000$
14. $760,000 \times 18,000$
15. $2,650,000 \times 41,000$

Find the square root of each of the following numbers.

16. 5,625
17. 729
18. 361
19. 3,969
20. 7,056
21. 10,816
22. 42,849
23. 121,801
24. 145,161
25. 574,564

26. Change the following incorrect statement to a correct one by moving a single pin.

$$\mathsf{I \div \vee II = I}$$

Fractions

6.1. Fraction. The concept of a fraction originated in ancient times. The earliest records of mathematical notations show that the Babylonians considered the unit as divided into sixty equal parts, expressing a fraction as approximately so many sixtieths. The more advanced Egyptians and Greeks expressed a fraction as the sum of several unit fractions. In any case, their numerals were not suited for easy and efficient handling. For example,* in Egyptian hieroglyphics, unit fractions were indicated by placing the symbol ⟅—⟆ over the figure representing the number of equal parts into which the unit had been divided. Thus, ⟅—⟆ denoted one-third; ⟅—⟆, one-fourth; ⟅—⟆, one-sixth; and so on. Special symbols were used for some fractions: ⟅ for one-half; ⟅ for two-thirds; and ⟅ for three-fourths. The Greeks used alphabetical numerals and wrote a fraction such as $\frac{43}{224}$, as

$$\zeta''\kappa\eta''\rho\iota\beta''\sigma\kappa\delta'' = \tfrac{1}{7} + \tfrac{1}{28} + \tfrac{1}{112} + \tfrac{1}{224}.$$

The absence of a simple and practical method of notation made operations with fractions very difficult for these ancient peoples.

The word *fraction*, from the Latin *frangere*, to break, literally indicates a fragment or part of a unit. The diagram, Fig. 21(a), shows a unit which has been divided into two equal parts; each

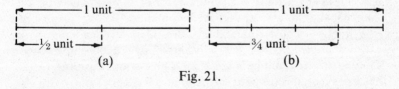

(a)   (b)

Fig. 21.

part is called a *half*, and one of these equal parts is the *fractional unit*. The same unit may be divided into four equal parts, Fig.

*Cajori, *op. cit.*

21(b); each part is then called a *fourth* and, in this case, the *fractional unit* is one-fourth. Three of these four equal parts comprise *three-fourths* of the whole unit. Hence, the concept of a fraction involves (1) a division of the unit, (2) a comparison of the part with the unit, and (3) a collection of the equal parts considered.

6.2. Common Fraction.   A *common fraction*, or simply a *fraction*, is defined as *one or more of the equal, integral parts of a unit*. Hence, it must be represented by two numbers: a number denoting the number of equal parts into which the unit is divided, and another number indicating the number of these equal parts which are taken.   These two numbers, called the *denominator* and the *numerator*, respectively, are the *terms* of a fraction.

The *denominator* indicates the number of equal, integral parts into which the unit is divided and is written under the *numerator*, which indicates the number of these equal parts which are taken. The two terms are separated by a horizontal or oblique line. Thus, in the fraction three-fourths, written $\frac{3}{4}$ or $\frac{3}{4}$, the denominator, 4, indicates that the unit has been divided into four equal parts and the numerator, 3, indicates that three of these equal parts are taken.

In other words, in the fraction $\frac{3}{4}$ or 3/4, the number 4, the denominator, directly indicates the number of equal parts of the unit and, indirectly, denominates or names the kind of things we are dealing with, that is, fourths. The number 3, the numerator, enumerates or counts the number of equal parts that are taken. Thus, if the fraction $\frac{3}{4}$ referred to a gallon, the denominator 4 tells us we are dealing with fourths of a gallon or with quarts, and the numerator 3 states we have taken three of these equal parts or quarts. So we have three-fourths, $\frac{3}{4}$, of a gallon or 3 quarts.

6.3. Types of Fractions.   The numerator of a fraction may be less than, equal to, or greater than its denominator.

A fraction whose numerator is less than its denominator is called a *proper fraction*.   For example, $\frac{3}{4}$ is a proper fraction. Thus, a proper fraction is less than a unit and, according to the basic concept, it is properly a fraction.

A fraction whose numerator is equal to its denominator is, by definition, equal to the unit.   For example, $\frac{4}{4}$ states that the unit has been divided into four equal parts and that four of these equal parts have been taken, that is, we have the whole unit.   Thus, $\frac{4}{4}$,

$\frac{5}{5}, \frac{6}{6}$, and so on, are all equal because each of these fractions equals one unit, so that $\frac{4}{4} = \frac{5}{5} = \frac{6}{6} = 1$.

A fraction whose numerator is greater than its denominator is called an *improper fraction*. Thus, $\frac{7}{4}$ is an improper fraction. Every improper fraction is greater than the unit. For example, $\frac{7}{4}$ is greater than the unit because each unit has been divided into 4 equal parts and we have taken 7 of these equal parts. Hence, $\frac{7}{4}$ exceeds the unit by $\frac{3}{4}$.

6.4. Mixed Numbers. An improper fraction expressed as a whole number and a fraction is called a *mixed number*. Taking the improper fraction $\frac{7}{4}$ as an example, seven-fourths of a unit is one whole unit and three-fourths of another equal unit, namely $1 + \frac{3}{4}$. A mixed number such as $1 + \frac{3}{4}$ is written $1\frac{3}{4}$.

6.5. Zero Denominator. As shown in Section 4.2, division by zero is excluded. Since a fraction, by definition, implies division and since division by zero is excluded, a fraction is *not defined*, that is, has no definite value, when its denominator is zero. Hence, the expression $\frac{12}{0}$ has no meaning.

EXERCISE 21

1. (a) How many eighths in one unit?   (b) In 4 units?   (c) In 6 units?
2. Write an even number using only odd digits.
3. (a) How many fourths are there in one half of a unit?   (b) How many eighths?   (c) How many tenths?
4. If a sum of money is divided into seven equal parts and a man receives four of these equal parts while another man receives the remainder, what are the shares that each man receives called?
5. Given the following fractions, $\frac{8}{15}, \frac{15}{9}, \frac{23}{23}, \frac{79}{78}, \frac{13}{28}, \frac{12}{12}, \frac{3}{5}$, state which (a) are greater than the unit; (b) are less than the unit; (c) are equal to the unit; (d) has the greatest value; (e) has the least value.
6. One boy has 5 chocolate bars and another boy has 3. A third boy comes along and they share the chocolate bars equally. If all the chocolate bars have the same value and the third boy pays 8 cents for his share, how should the 8 cents be divided?
7. How much must be added to each of the following fractions to make each fraction equal to a unit: (a) $\frac{5}{7}$;   (b) $\frac{7}{11}$;   (c) $\frac{13}{25}$;   (d) $\frac{14}{19}$?
8. How much larger than a unit are each of the following fractions: (a) $\frac{7}{5}$;   (b) $\frac{11}{8}$;   (c) $\frac{25}{13}$;   (d) $\frac{61}{8}$?

9. If 2 is added to each of the numerators of the following fractions, by how much does the fraction increase: (a) $\frac{5}{7}$; (b) $\frac{1}{5}$; (c) $\frac{3}{4}$; (d) $\frac{7}{8}$?

10. If 3 is subtracted from each of the numerators of the following fractions, by how much does the fraction decrease: (a) $\frac{25}{13}$; (b) $\frac{7}{9}$; (c) $\frac{5}{18}$; (d) $\frac{4}{7}$?

6.6. Properties of Fractions. The fraction $\frac{4}{15}$ represents 4 of the 15 equal parts into which the unit has been divided. If the numerator 4 is multiplied by any number, such as 2, the resulting fraction $\frac{8}{15}$ contains 8 of the 15 equal parts, that is, twice as many equal parts. Hence, $\frac{8}{15}$ is twice $\frac{4}{15}$. It is easy to see that this result holds for any fraction, for multiplying only the numerator by a number increases the number of equal parts taken by the number of times that the multiplier represents. Therefore,

1) *Multiplying only the numerator of a fraction by a number multiplies the fraction by that number.*

If the numerator 4 of the fraction $\frac{4}{15}$ is divided by any number, say 2, the resulting fraction $\frac{2}{15}$ contains 2 of the 15 equal parts, that is, half as many equal parts. Hence, $\frac{2}{15}$ is half of $\frac{4}{15}$. In general, dividing only the numerator of a fraction by a number decreases the number of parts taken by the number of times represented by the divisor. Hence,

2) *Dividing only the numerator of a fraction by a number divides the fraction by that number.*

If only the denominator of $\frac{4}{15}$ is multiplied by 3, the unit in the resulting fraction, $\frac{4}{15 \times 3} = \frac{4}{45}$, will be divided into 45 equal parts, that is, into three times as many equal parts. Hence, each part will be one-third as large as it was, and since the same number of equal parts, 4, is taken, the value of the resulting fraction $\frac{4}{45}$ will be one-third as large as it was. For if the denominator of a fraction is multiplied by any number, the unit in the resulting fraction will be divided into that many more number of times. Hence, each equal part of the resulting fraction is decreased by that number of times. Now since the number of parts taken remains the same, the value of the resulting fraction is decreased by that number of times. Therefore,

3) *Multiplying only the denominator of a fraction by a number divides the fraction by that number.*

If the denominator 15 of the fraction $\frac{4}{15}$ is divided by 3, the unit in the resulting fraction, $\frac{4}{15 \div 3} = \frac{4}{5}$, will be divided into 5 equal parts, that is, into one-third as many equal parts. Hence, each equal part will be three times as large as it was, and since the same number of equal parts, 4, is taken, the value of the resulting fraction $\frac{4}{5}$ will be three times as large as it was, that is, $\frac{4}{5}$ is three times $\frac{4}{15}$. If only the denominator of a fraction is divided by any number, the number of equal parts in the resulting unit will be decreased by that number of times. Hence, each equal part in the resulting fraction will be that many times larger. Since the same number of equal parts are taken, the resulting fraction will be that number of times larger. Hence,

4) *Dividing only the denominator of a fraction by a number multiplies the fraction by that number.*

Multiplying the numerator of a fraction by a number multiplies the fraction by that number (property 1), and multiplying the denominator of this fraction by that same number divides the fraction by that same number (property 3). Hence, the operations cancel each other and the fraction remains unchanged. Moreover, dividing the numerator by any number divides the fraction by this number (property 2), and dividing the denominator by the same number multiplies the fraction by this number (property 4). Again the operations cancel each other and the fraction remains unchanged. Therefore,

5) *Multiplying or dividing the numerator and the denominator of a fraction by the same number (except zero) does not change the value of the fraction.*

For example, $\frac{1}{2} = \frac{1 \times 4}{2 \times 4} = \frac{4}{8}$ for if we take 4 of 8 equal parts, we have taken one-half of the unit. Likewise, since $\frac{4}{8}$ is four of eight equal parts or one-half of the unit, $\frac{4 \div 4}{8 \div 4} = \frac{1}{2}$.

EXERCISE 22

1. What are the changes that occur in the fraction $\frac{3}{11}$ if the numerator is (a) multiplied by 2? (b) divided by 3?
2. What are the changes that occur in the fraction $\frac{8}{21}$ if (a) 32 is substituted for 8? (b) 2 is substituted for 8?

3. What are the changes that occur in the fraction $\frac{5}{12}$ if the denominator is (a) multiplied by 2?   (b) divided by 3?

4. What are the changes that occur in the fraction $\frac{4}{21}$ if (a) 7 is substituted for 21?   (b) 63 is substituted for 21?

5. What are the changes that occur in the fraction $\frac{6}{10}$ if both terms are (a) multiplied by 3?   (b) divided by 2?

6. $\dfrac{8}{12} = \dfrac{4}{?}$ 8. $\dfrac{18}{27} = \dfrac{6}{?}$ 10. $\dfrac{8}{10} = \dfrac{4}{?}$ 12. $\dfrac{5}{7} = \dfrac{10}{?}$

7. $\dfrac{5}{15} = \dfrac{1}{?}$ 9. $\dfrac{7}{14} = \dfrac{?}{2}$ 11. $\dfrac{6}{10} = \dfrac{?}{5}$ 13. $\dfrac{2}{3} = \dfrac{?}{12}$

14. Using two methods, multiply the fractions $\dfrac{3}{14}, \dfrac{2}{21}, \dfrac{5}{35}$ by 7.

15. Using two methods, divide the fractions $\dfrac{6}{7}, \dfrac{12}{17}, \dfrac{15}{28}$ by 3.

6.7. Reduction of Fractions.

In many cases fractions must be changed from one form to another. Any process which changes a fraction from one form to another, without changing its value, is called *reduction*. The following paragraphs present the most important cases of reduction.

6.7–1. Reduction of Mixed Numbers and Improper Fractions.

As already stated in Section 6.4, a mixed number, such as $4\frac{2}{3}$, means $4 + \frac{2}{3}$ or 4 units plus 2 thirds of a unit. Here, the unit contains 3 thirds, hence 4 units contain $4 \times 3 = 12$ thirds. But 12 thirds plus 2 thirds equals 14 thirds; hence, $4\frac{2}{3} = \frac{14}{3}$. Note that $\frac{14}{3} = \frac{4 \times 3 + 2}{3}$. Therefore,

1) *To change a mixed number to an improper fraction, multiply the whole number by the denominator of the fraction. Add this product to the numerator, placing the sum over the denominator.*

Example.   Reduce $5\frac{1}{3}$ to an improper fraction.

Solution. $5\dfrac{1}{3} = \dfrac{5 \times 3 + 1}{3} = \dfrac{16}{3}.$ *Ans.*

An improper fraction such as $\frac{17}{5}$ indicates that after dividing each of a number of equal units into 5 equal parts, 17 of these equal parts were taken. Since 5 fifths equals one unit, there are as many whole units in 17 fifths as 5 is contained in 17. Hence, there are 3 units and 2 fifths of a unit in $\frac{17}{5}$; that is, $\dfrac{17}{5} = 3 + \dfrac{2}{5} = 3\dfrac{2}{5}.$

Note that if 17 is divided by 5 we obtain a quotient of 3 and a remainder of 2. Therefore,

2) *To change an improper fraction to a mixed number, divide the numerator by the denominator. The quotient is the whole number, and the remainder placed over the denominator is the fraction.*

Example. Reduce $\frac{31}{4}$ to a mixed number.

Solution.

$$4\overline{)31} \quad\quad \text{Hence, } \frac{31}{4} = 7\frac{3}{4}. \quad \textit{Ans.}$$
$$\underline{28}$$
$$3$$

quotient 7.

EXERCISE 23

Reduce each of the following mixed numbers to an improper fraction.

1. $1\frac{1}{2}$ 3. $2\frac{1}{4}$ 5. $4\frac{2}{3}$ 7. $6\frac{5}{6}$ 9. $4\frac{5}{7}$ 11. $15\frac{3}{4}$

2. $1\frac{1}{5}$ 4. $6\frac{1}{5}$ 6. $8\frac{3}{4}$ 8. $3\frac{3}{8}$ 10. $12\frac{2}{3}$ 12. $11\frac{3}{7}$

Reduce each of the following improper fractions to a mixed number.

13. $\frac{7}{2}$ 15. $\frac{25}{4}$ 17. $\frac{85}{9}$ 19. $\frac{112}{11}$ 21. $\frac{43}{13}$ 23. $\frac{910}{17}$

14. $\frac{14}{3}$ 16. $\frac{48}{5}$ 18. $\frac{62}{5}$ 20. $\frac{89}{14}$ 22. $\frac{103}{12}$ 24. $\frac{1,054}{25}$

25. Write 2 using exactly seven twos.

6.7–2. Reduction to Higher Terms.   Assume that a fraction, say $\frac{2}{3}$, is to be expressed as an equivalent fraction whose denominator is 18. One unit equals $\frac{18}{18}$; hence, $\frac{1}{3}$ of a unit equals $\frac{1}{3}$ of 18 eighteenths or 6 eighteenths. Then $\frac{2}{3}$ equals twice 6 eighteenths or 12 eighteenths. Hence, $\frac{2}{3} = \frac{12}{18}$. Note that we actually multiply the numerator of $\frac{2}{3}$ by the number 6, which indicates how many times 3, the given denominator, is contained in the required denominator 18. That is, we multiply both terms of the given fraction $\frac{2}{3}$ by the same number 6 and hence, by property 5, page 101, we do not change the value of the fraction. Therefore,

To reduce a fraction to higher terms:

1) *Divide the new denominator by the denominator of the given fraction.*

2) *Multiply the numerator and the denominator of the given fraction by the quotient obtained in 1.*

Example. Reduce $\frac{3}{4}$ to an equivalent fraction having 24 as a denominator.

Solution. $24 \div 4 = 6.$

$$\frac{3}{4} = \frac{3 \times 6}{4 \times 6} = \frac{18}{24}. \quad Ans.$$

6.7–3. Reduction to Lower Terms. Now suppose that the fraction $\frac{12}{15}$ is to be expressed as an equivalent fraction whose denominator is 5. One unit is $\frac{15}{15}$; hence $\frac{1}{5}$ of a unit equals $\frac{1}{5}$ of 15 fifteenths or 3 fifteenths. Then, $\frac{1}{3}$ of the number of fifteenths must equal the number of fifths; and since $\frac{1}{3}$ of 12 is 4, then $\frac{12}{15} = \frac{4}{5}$. Note that we divide the numerator of $\frac{12}{15}$ by the number 3, which indicates how many times the denominator 5 is contained in the denominator, 15, of the given fraction. That is, we divide both terms of the given fraction $\frac{12}{15}$ by the same number 3 and hence, by property 5, page 101, we do not change the value of the fraction. Therefore,

To reduce a fraction to lower terms:

1) Divide the denominator of the given fraction by the new denominator.

2) Divide the numerator and the denominator of the given fraction by the quotient obtained in 1.

Example. Reduce $\frac{84}{128}$ to an equivalent fraction having 32 as a denominator.

Solution. $128 \div 32 = 4.$

$$\frac{84}{128} = \frac{84 \div 4}{128 \div 4} = \frac{21}{32}. \quad Ans.$$

6.7–4. Reduction to Lowest Terms. A fraction is said to be in its *lowest terms* or in its *simplest form* when both numerator and denominator are integral numbers which have no common factor except 1. When both terms of a fraction contain one or more common factors, the fraction may be reduced to its lowest terms by dividing both terms by the common factor or factors (property 5, page 101). The process of dividing both terms of a fraction by a common factor is called *cancelling*.

Example. Reduce $\frac{30}{105}$ to its lowest terms.

Solution. $\dfrac{30}{105} = \dfrac{\overset{1}{\cancel{3}} \times \overset{1}{\cancel{5}} \times 2}{\underset{1}{\cancel{3}} \times \underset{1}{\cancel{5}} \times 7} = \dfrac{2}{7}. \quad Ans.$

6.7–5. Reduction of Compound Expressions. Some of the calculations involved in a series of indicated operations may be eliminated by first cancelling all common factors.

Example. Simplify: $(3 \times 34) \div (17 \times 9)$.

Solution. $(3 \times 34) \div (17 \times 9) = \dfrac{\overset{1}{\cancel{3}} \times \overset{2}{\cancel{34}}}{\underset{1}{\cancel{17}} \times \underset{3}{\cancel{9}}} = \dfrac{2}{3}$. *Ans.*

Note that only the common factors of the numerator and the denominator may be cancelled. For example, $\dfrac{2 + 5}{2 + 6}$ is *not* equal to $\dfrac{\overset{1}{\cancel{2}} + 5}{\underset{1}{\cancel{2}} + 6}$ or $\dfrac{6}{7}$ for 2 is not a common factor of the numerator and the denominator. Actually, $\dfrac{2 + 5}{2 + 6} = \dfrac{7}{8}$. Moreover, $\dfrac{5 + 3}{15}$ is *not* equal to $\dfrac{\overset{1}{\cancel{5}} + \overset{1}{\cancel{3}}}{\underset{\underset{1}{\cancel{3}}}{\cancel{15}}}$ or 2, for neither 5 nor 3 is a common factor of the numerator; in fact, $\dfrac{5 + 3}{15} = \dfrac{8}{15}$.

EXERCISE 24

Reduce each of the following fractions to their indicated higher or lower terms.

1. $\dfrac{1}{5} = \dfrac{?}{20}$ 5. $\dfrac{6}{7} = \dfrac{?}{63}$ 9. $\dfrac{6}{18} = \dfrac{?}{9}$ 13. $\dfrac{15}{90} = \dfrac{?}{18}$

2. $\dfrac{2}{3} = \dfrac{?}{12}$ 6. $\dfrac{8}{29} = \dfrac{?}{174}$ 10. $\dfrac{6}{27} = \dfrac{?}{9}$ 14. $\dfrac{52}{816} = \dfrac{?}{204}$

3. $\dfrac{2}{13} = \dfrac{?}{39}$ 7. $\dfrac{23}{25} = \dfrac{?}{200}$ 11. $\dfrac{14}{34} = \dfrac{?}{17}$ 15. $\dfrac{119}{364} = \dfrac{?}{52}$

4. $\dfrac{8}{17} = \dfrac{?}{102}$ 8. $\dfrac{71}{83} = \dfrac{?}{415}$ 12. $\dfrac{32}{24} = \dfrac{?}{3}$ 16. $\dfrac{77}{595} = \dfrac{?}{85}$

Reduce each of the following fractions to their lowest terms.

17. $\dfrac{6}{15}$ 19. $\dfrac{6}{21}$ 21. $\dfrac{14}{21}$ 23. $\dfrac{70}{126}$ 25. $\dfrac{99}{165}$ 27. $\dfrac{343}{539}$

18. $\dfrac{9}{12}$ **20.** $\dfrac{12}{44}$ **22.** $\dfrac{30}{42}$ **24.** $\dfrac{28}{36}$ **26.** $\dfrac{84}{126}$ **28.** $\dfrac{121}{143}$

Simplify by cancelling.

29. $(3 \times 2 \times 5) \div (6 \times 4 \times 10)$
30. $(2 \times 5 \times 6) \div (14 \times 8 \times 10)$
31. $(21 \times 55) \div (15 \times 77 \times 13)$
32. $(32 \times 49 \times 6) \div (21 \times 16 \times 70)$
33. $(8 \times 9 \times 49 \times 33) \div (21 \times 28 \times 11 \times 6)$
34. $(5 \times 20 \times 18 \times 7) \div (3 \times 6 \times 10 \times 21)$

6.8. Lowest Common Denominator. Some operations with fractions require that the given fractions be changed to equivalent fractions having the same or a common denominator. A *common denominator* of two or more fractions is any number which contains all the given denominators as factors, that is, a common multiple of all the denominators (see Section 5.6). In the majority of cases, a lowest common denominator should be obtained. The *lowest common denominator* (L. C. D.) of two or more fractions is the smallest number which contains all the denominators as factors. In other words, the L. C. D. of two or more fractions is the lowest common multiple (L. C. M.) of all the denominators. For example, in the fractions $\frac{1}{2}$, $\frac{2}{3}$, and $\frac{5}{6}$, the number 12 is a common denominator, for it is a common multiple of all the denominators and hence, contains all of them as factors. But, 6 is the lowest common denominator because it is the least common multiple of all the denominators and the smallest number that contains all of them as factors. Usually, the L. C. D. can be found by inspection. In other cases, use any of the methods shown in Chapter 5 to find the L. C. M.

Example. Find the L. C. D. of $\frac{1}{4}$, $\frac{5}{6}$, $\frac{4}{15}$, and $\frac{11}{24}$.
Solution. Find the L. C. M. of 4, 6, 15, and 24. As shown at the left, the L. C. M. of 4, 6, 15, and 24 is $3 \times 4 \times 2 \times 5 = 120$.

3	4	6	15	24
4	4	2	5	8
2	1	2	5	2
	1	1	5	1

Therefore, the L. C. D. = 120. *Ans.*

EXERCISE 25

Find the L. C. D. of each of the following.

1. $\dfrac{2}{7}$, $\dfrac{4}{21}$ **7.** $\dfrac{1}{3}$, $\dfrac{4}{9}$, $\dfrac{5}{27}$, $\dfrac{2}{81}$ **13.** $\dfrac{3}{16}$, $\dfrac{2}{21}$, $\dfrac{7}{15}$, $\dfrac{11}{48}$

2. $\dfrac{2}{5}, \dfrac{4}{35}$ **8.** $\dfrac{5}{6}, \dfrac{1}{9}, \dfrac{7}{12}, \dfrac{5}{36}$ **14.** $\dfrac{1}{5}, \dfrac{5}{12}, \dfrac{3}{8}, \dfrac{11}{120}$

3. $\dfrac{5}{6}, \dfrac{3}{10}$ **9.** $\dfrac{1}{6}, \dfrac{2}{9}, \dfrac{5}{8}$ **15.** $\dfrac{1}{2}, \dfrac{2}{9}, \dfrac{5}{12}, \dfrac{7}{24}$

4. $\dfrac{2}{3}, \dfrac{5}{6}, \dfrac{5}{24}$ **10.** $\dfrac{3}{10}, \dfrac{2}{27}, \dfrac{7}{30}$ **16.** $\dfrac{3}{40}, \dfrac{7}{60}, \dfrac{5}{18}, \dfrac{13}{180}$

5. $\dfrac{1}{3}, \dfrac{2}{9}, \dfrac{7}{18}$ **11.** $\dfrac{1}{6}, \dfrac{9}{20}, \dfrac{4}{25}$ **17.** $\dfrac{3}{16}, \dfrac{7}{72}, \dfrac{5}{49}, \dfrac{11}{32}, \dfrac{1}{10}$

6. $\dfrac{5}{6}, \dfrac{1}{12}, \dfrac{7}{24}$ **12.** $\dfrac{2}{15}, \dfrac{4}{45}, \dfrac{7}{60}$ **18.** $\dfrac{7}{91}, \dfrac{24}{130}, \dfrac{6}{70}, \dfrac{9}{160}$

6.9. Addition of Fractions. The denominator of a fraction names the kind of things with which we are dealing, for example, fourths or fifths. If the quantities to be added are of the same denomination, no difficulty is encountered. Thus, 1 quart added to 3 quarts gives as a result 4 quarts or 1 gallon. Similarly, 1 fourth added to 3 fourths gives as a result 4 fourths or 1 unit; that is, $\frac{1}{4} + \frac{3}{4} = \frac{4}{4} = 1$. Hence,

To add fractions with the same denominator add the numerators and place the result over the common denominator.

Example. Add $\frac{2}{9}$ to $\frac{5}{9}$.

Solution. $\dfrac{2}{9} + \dfrac{5}{9} = \dfrac{2+5}{9} = \dfrac{7}{9}$. *Ans.*

When the quantities to be added are of different denominations they must be changed to a common denominator before a sum can be obtained. Thus, if 4 inches is to be added to 2 feet, either the 4 inches must be expressed in feet or the 2 feet must be expressed in inches. Similarly, if $\frac{1}{3}$ is to be added to $\frac{2}{5}$, it is first necessary to change these fractions to equivalent fractions having the same, or a common, denominator so that these fractions can then be added. Therefore,

To add fractions with different denominators:
1) *Find the L. C. D. of the fractions.*
2) *Change the fractions to equivalent fractions having the lowest common denominator.*
3) *Add the numerators, placing the result over the lowest common denominator.*
4) *Express the resulting fraction in its lowest terms.*

Example. Add $\frac{1}{3}$ to $\frac{2}{5}$.

Solution. L. C. D. $= 3 \times 5 = 15$.

$$\frac{1}{3} = \frac{1 \times 5}{3 \times 5} = \frac{5}{15}; \qquad \frac{2}{5} = \frac{2 \times 3}{5 \times 3} = \frac{6}{15};$$

$$\frac{1}{3} + \frac{2}{5} = \frac{5}{15} + \frac{6}{15} = \frac{5 + 6}{15} = \frac{11}{15}. \quad \textit{Ans.}$$

Example. Add $\frac{1}{6}$, $\frac{7}{20}$, and $\frac{4}{25}$.

Solution.

$$\begin{array}{r|ccc} 5 & 6 & 20 & 25 \\ 2 & 6 & 4 & 5 \\ \hline & 3 & 2 & 5 \end{array}$$

L. C. D. $= 5 \times 2 \times 3 \times 2 \times 5 = 300$.

$$\frac{1}{6} = \frac{1 \times 50}{6 \times 50} = \frac{50}{300}; \frac{7}{20} = \frac{7 \times 15}{20 \times 15} = \frac{105}{300}; \frac{4}{25} = \frac{4 \times 12}{25 \times 12} = \frac{48}{300};$$

$$\frac{1}{6} + \frac{7}{20} + \frac{4}{25} = \frac{50}{300} + \frac{105}{300} + \frac{48}{300} = \frac{50 + 105 + 48}{300} = \frac{203}{300}. \quad \textit{Ans.}$$

Example. Add $3\frac{3}{4}$ to $5\frac{5}{9}$.

Solution. $3\frac{3}{4} = \frac{15}{4};\quad 5\frac{5}{9} = \frac{50}{9};$

$$3\frac{3}{4} + 5\frac{5}{9} = \frac{15}{4} + \frac{50}{9}. \quad \text{L. C. D.} = 36.$$

$$3\frac{3}{4} + 5\frac{5}{9} = \frac{15}{4} + \frac{50}{9} = \frac{(15 \times 9) + (50 \times 4)}{36}$$

$$= \frac{135 + 200}{36} = \frac{335}{36} = 9\frac{11}{36}. \quad \textit{Ans.}$$

Mixed numbers may also be added as follows: To the sum of the integers added separately, add the sum of the fractions.

Example. Add $3\frac{3}{4}$ to $5\frac{5}{9}$.

Solution. $3 + 5 = 8$;

$$\frac{3}{4} + \frac{5}{9} = \frac{(3 \times 9) + (5 \times 4)}{36} = \frac{27 + 20}{36} = \frac{47}{36} = 1\frac{11}{36};$$

$$8 + 1\frac{11}{36} = 9\frac{11}{36}. \quad \textit{Ans.}$$

Example. Perform the indicated operation: $4 + \frac{1}{6} + 3\frac{3}{10}$.

Solution. $4 + \frac{1}{6} + 3\frac{3}{10} = 4 + \frac{1}{6} + \frac{33}{10}. \quad \text{L. C. D.} = 30.$

$$4 + \frac{1}{6} + \frac{33}{10} = \frac{(4 \times 30) + (1 \times 5) + (33 \times 3)}{30} = \frac{120 + 5 + 99}{30}$$

$$= \frac{224}{30} = \frac{112}{15} = 7\frac{7}{15}. \quad \textit{Ans.}$$

Or

$$4 + 3 = 7; \quad \frac{1}{6} + \frac{3}{10} = \frac{(1 \times 5) + (3 \times 3)}{30} = \frac{5 + 9}{30} = \frac{14}{30} = \frac{7}{15};$$

$$7 + \frac{7}{15} = 7\frac{7}{15}. \quad Ans.$$

6.10. Subtraction of Fractions. If the quantities to be subtracted are of the same denomination, they can be subtracted by finding the difference between the numerators. Thus, 1 quart subtracted from 3 quarts gives as a result 2 quarts. Similarly, 1 fourth subtracted from 3 fourths gives as a result 2 fourths or one-half of a unit; that is, $\frac{3}{4} - \frac{1}{4} = \frac{3-1}{4} = \frac{2}{4} = \frac{1}{2}$. Hence,

To subtract fractions with the same denominator find the difference between the numerators and place the result over the common denominator.

Example. Subtract $\frac{3}{7}$ from $\frac{5}{7}$.

Solution. $\frac{5}{7} - \frac{3}{7} = \frac{5-3}{7} = \frac{2}{7}. \quad Ans.$

As in the addition of fractions, if the quantities to be subtracted are of different denominations, they must be changed to a common denominator before a difference can be obtained. Thus, if 4 inches is to be subtracted from 2 feet, either the 4 inches must be expressed in feet or the 2 feet must be expressed in inches. Similarly, if $\frac{1}{3}$ is to be subtracted from $\frac{2}{5}$, it is first necessary to change these fractions to equivalent fractions having the same, or a common, denominator. Therefore,

To subtract fractions with different denominators:
1) *Find the L. C. D. of the fractions.*
2) *Change the fractions to equivalent fractions having the lowest common denominator.*
3) *Subtract the numerators, placing the result over the lowest common denominator.*
4) *Express the resulting fraction in its lowest terms.*

Example. Subtract $\frac{1}{3}$ from $\frac{2}{5}$.

Solution. L. C. D. $= 3 \times 5 = 15$.

$$\frac{1}{3} = \frac{1 \times 5}{3 \times 5} = \frac{5}{15}; \qquad \frac{2}{5} = \frac{2 \times 3}{5 \times 3} = \frac{6}{15};$$

$$\frac{2}{5} - \frac{1}{3} = \frac{6}{15} - \frac{5}{15} = \frac{6-5}{15} = \frac{1}{15}. \quad Ans.$$

Example. Subtract $7\frac{3}{4}$ from $15\frac{5}{9}$.

Solution. $7\frac{3}{4} = \frac{31}{4}$; $\quad 15\frac{5}{9} = \frac{140}{9}$; $\quad 15\frac{5}{9} - 7\frac{3}{4} = \frac{140}{9} - \frac{31}{4}$.

L. C. D. = 36.

$$\frac{140}{9} - \frac{31}{4} = \frac{(140 \times 4) - (31 \times 9)}{36}$$

$$= \frac{560 - 279}{36} = \frac{281}{36} = 7\frac{29}{36}. \quad Ans.$$

Example. Perform the indicated operations: $\frac{3}{4} + \frac{2}{5} - \frac{5}{6}$.

Solution. L. C. D. = $4 \times 5 \times 3 = 60$.

$$\frac{3}{4} = \frac{3 \times 15}{4 \times 15} = \frac{45}{60}; \quad \frac{2}{5} = \frac{2 \times 12}{5 \times 12} = \frac{24}{60}; \quad \frac{5}{6} = \frac{5 \times 10}{6 \times 10} = \frac{50}{60};$$

$$\frac{45}{60} + \frac{24}{60} - \frac{50}{60} = \frac{45 + 24 - 50}{60} = \frac{19}{60}. \quad Ans.$$

EXERCISE 26

Perform the indicated operations.

1. $\dfrac{1}{2} + \dfrac{1}{3}$ **2.** $\dfrac{3}{5} - \dfrac{1}{5}$ **3.** $\dfrac{5}{9} + \dfrac{1}{3}$

4. $\dfrac{5}{12} + \dfrac{3}{4}$ **5.** $\dfrac{4}{5} - \dfrac{3}{4}$ **6.** $\dfrac{5}{26} + \dfrac{2}{39}$

7. $\dfrac{5}{6} - \dfrac{4}{15}$ **8.** $16 - \dfrac{5}{11}$ **9.** $5\dfrac{3}{7} + 3\dfrac{5}{14}$

10. $9\dfrac{2}{7} - 5\dfrac{3}{4}$ **11.** $7\dfrac{3}{5} + 3\dfrac{1}{10}$ **12.** $8\dfrac{5}{6} - 5\dfrac{1}{12}$

13. $3\dfrac{9}{55} + 5\dfrac{13}{44}$ **14.** $9\dfrac{7}{8} - 2\dfrac{5}{24}$ **15.** $\dfrac{3}{4} + \dfrac{2}{3} + \dfrac{8}{9}$

16. $\dfrac{2}{5} - \dfrac{1}{8} - \dfrac{1}{6}$ **17.** $\dfrac{2}{7} + \dfrac{3}{4} + \dfrac{1}{2}$ **18.** $\dfrac{2}{3} - \dfrac{1}{7} - \dfrac{2}{9}$

19. $\dfrac{3}{4} + \dfrac{2}{3} - \dfrac{5}{6}$ **20.** $\dfrac{3}{4} - \dfrac{5}{8} + \dfrac{7}{12}$ **21.** $\dfrac{7}{9} + \dfrac{16}{25} - \dfrac{8}{15}$

22. $2\dfrac{15}{36} + \dfrac{9}{60} + \dfrac{25}{72}$ **23.** $4 + 1\dfrac{1}{3} - \dfrac{2}{5}$

24. $3\dfrac{3}{35} + 1\dfrac{2}{105} + 2\dfrac{11}{420}$ **25.** $9 - 2\dfrac{1}{8} - 3\dfrac{5}{24}$

26. $\dfrac{15}{56} - \dfrac{7}{24} + \dfrac{21}{40} - \dfrac{5}{48}$ **27.** $\dfrac{7}{45} + \dfrac{5}{54} + \dfrac{15}{36} + \dfrac{11}{63}$

28. $\dfrac{9}{11} - \dfrac{7}{22} - \dfrac{21}{44} + \dfrac{16}{55}$ **29.** $5\dfrac{1}{5} - \dfrac{5}{8} + \dfrac{7}{40} - 3$

30. $2\dfrac{3}{7} + 4\dfrac{1}{6} + 5\dfrac{1}{2} + 1\dfrac{3}{14}$ **31.** $4\dfrac{1}{19} - 3\dfrac{3}{8} + 7\dfrac{1}{16} - \dfrac{1}{2}$

32. Arrange the digits from 0 to 9 so that the sum is equal to one.

33. A turkey weighs 10 pounds and a half of its weight besides. What is its weight?

34. A firm received the following consignments of lard: $54\frac{2}{3}$ lb., $28\frac{2}{5}$ lb., $16\frac{1}{2}$ lb., $36\frac{3}{4}$ lb., and $64\frac{3}{8}$ lb. They sold $64\frac{5}{8}$ lb., $27\frac{5}{8}$ lb., and $58\frac{2}{8}$ lb. (a) How many pounds did they receive? (b) How many pounds did they sell? (c) How many pounds were left?

35. A farmer had three corn fields. The first produced $254\frac{3}{8}$ bushels, the second, $179\frac{9}{10}$, and the third, $237\frac{1}{4}$. He sold $275\frac{1}{2}$ bushels and sent $189\frac{5}{8}$ bushels to the mill. How many bushels did he have left?

6.11. Multiplication of Fractions.

Multiplication is a special case of addition in which the numbers added are all equal (Section 3.1). Thus, multiplication is fundamentally a short method of repeated addition. To multiply $\frac{2}{11}$ by 3 is equivalent to $\dfrac{2}{11} + \dfrac{2}{11} + \dfrac{2}{11} = \dfrac{2+2+2}{11} = \dfrac{2 \times 3}{11} = \dfrac{6}{11}$. Thus,

To multiply a whole number by a fraction or a fraction by a whole number, multiply the numerator of the fraction by the whole number placing the product over the denominator.

Example. Multiply 3 by $\frac{2}{7}$.

Solution. $3 \times \dfrac{2}{7} = \dfrac{3 \times 2}{7} = \dfrac{6}{7}$. *Ans.*

The general rule for the multiplication of a fraction by another fraction may be derived from the following analysis. Find the product of $\frac{2}{5} \times \frac{3}{7}$; or, what is $\frac{2}{5}$ of $\frac{3}{7}$? Note that $\frac{1}{5}$ of $\frac{1}{7}$ is one of the five equal, integral parts into which $\frac{1}{7}$ can be divided. If each seventh is divided into five equal parts, $\frac{7}{7}$, or the unit, will be divided into 5×7 or 35 equal parts, and each part will be $\frac{1}{35}$ of a unit. Hence, $\frac{1}{5}$ of $\frac{1}{7}$ is $\frac{1}{35}$; that is, $\frac{1}{5} \times \frac{1}{7} = \frac{1}{35}$. It follows that $\frac{1}{5}$ of $\frac{3}{7}$ is $3 \times \dfrac{1}{35} = \dfrac{3 \times 1}{35} = \dfrac{3}{35}$ and $\frac{2}{5}$ of $\frac{3}{7}$ is $2 \times \dfrac{3}{35} = \dfrac{2 \times 3}{35} = \dfrac{6}{35}$.

If this solution is examined carefully, it may be seen that the numerators have been multiplied together and the denominators have also been multiplied together. Hence,

To multiply a fraction by another fraction, place the product of the numerators over the product of the denominators.

Example.   Find the product of $\frac{2}{9} \times \frac{3}{4}$.

Solution. $\frac{2}{9} \times \frac{3}{4} = \frac{2 \times 3}{9 \times 4} = \frac{6}{36} = \frac{1}{6}$.   *Ans.*

Note that since the product of the numerators is the numerator of the product and the product of the denominators is the denominator of the product, that is, each of the numerators is a factor of the numerator of the product and each of the denominators is a factor of the denominator of the product, the work may be simplified by cancelling. Thus,

$$\frac{2}{9} \times \frac{3}{4} = \frac{\overset{1}{2} \times \overset{1}{3}}{\underset{3}{9} \times \underset{2}{4}} = \frac{1}{6}. \textit{Ans.}$$

Example.   Find the product of $\frac{5}{9} \times \frac{3}{4} \times \frac{2}{35}$.

Solution. $\frac{5}{9} \times \frac{3}{4} \times \frac{2}{35} = \frac{\overset{1}{5} \times \overset{1}{3} \times \overset{1}{2}}{\underset{3}{9} \times \underset{2}{4} \times \underset{7}{35}} = \frac{1}{42}$.   *Ans.*

Example.   Find the product of $5 \times \frac{5}{34} \times 2\frac{3}{7}$.

Solution. $5 \times \frac{5}{34} \times 2\frac{3}{7} = \frac{5 \times 5 \times \overset{1}{17}}{\underset{2}{34} \times 7} = \frac{25}{14} = 1\frac{11}{14}$.   *Ans.*

Example.   A man owns $\frac{2}{5}$ of the total stock of a business and sells $\frac{3}{10}$ of his share. What part of the total stock does he still own?

Solution. $\frac{1}{10}$ of $\frac{1}{5}$ is one of the 10 equal, integral parts into which $\frac{1}{5}$ can be divided. If each fifth is divided into 10 equal parts, $\frac{5}{5}$, or the total stock, will be divided into 10×5 or 50 equal parts and each part will be $\frac{1}{50}$ of the total stock. It follows that $\frac{1}{10}$ of $\frac{2}{5}$ is $\frac{2}{50}$ and $\frac{3}{10}$ of $\frac{2}{5}$ is $3 \times \frac{2}{50} = \frac{6}{50} = \frac{3}{25}$ of the total stock. Thus, the man has sold $\frac{3}{25}$ of the total stock. Since he originally owned $\frac{2}{5}$ of the total stock, he now owns $\frac{2}{5} - \frac{3}{25} = \frac{(2 \times 5) - 3}{25} = \frac{10 - 3}{25} = \frac{7}{25}$.   *Ans.*

or

$\frac{\overset{1}{2}}{5} \times \frac{3}{\underset{5}{10}} = \frac{3}{25}$; $\frac{2}{5} - \frac{3}{25} = \frac{(2 \times 5) - 3}{25} = \frac{10 - 3}{25} = \frac{7}{25}$.   *Ans.*

EXERCISE 27

Perform the indicated operations.

1. $\frac{1}{2} \times \frac{2}{3}$ 2. $\frac{5}{6} \times \frac{3}{7}$ 3. $\frac{16}{19} \times \frac{5}{8}$

4. $\dfrac{1}{2} \times \dfrac{2}{3} \times \dfrac{3}{4}$ 5. $\dfrac{7}{10} \times \dfrac{5}{21} \times \dfrac{12}{13}$ 6. $\dfrac{6}{7} \times \dfrac{7}{8} \times \dfrac{8}{9}$

7. $\dfrac{13}{19} \times \dfrac{7}{13} \times \dfrac{19}{21}$ 8. $\dfrac{34}{51} \times \dfrac{19}{108} \times \dfrac{27}{38}$ 9. $3\dfrac{1}{4} \times 2\dfrac{2}{13}$

10. $2\dfrac{1}{2} \times \dfrac{1}{5} \times \dfrac{2}{7}$ 11. $1\dfrac{2}{7} \times \dfrac{5}{9} \times \dfrac{14}{17}$ 12. $\left(1 - \dfrac{3}{8}\right) \times 1\dfrac{3}{5}$

13. $\left(\dfrac{5}{8} + \dfrac{2}{9}\right) \times \left(\dfrac{2}{3} - \dfrac{50}{183}\right)$

14. $\left(\dfrac{11}{45} - \dfrac{1}{6}\right) \times \left(28\dfrac{1}{4} + 1\dfrac{3}{4}\right)$

15. $17\dfrac{3}{5} \times \left(4\dfrac{1}{6} - 2\dfrac{1}{2} + 1\dfrac{2}{3}\right) \times \dfrac{1}{176}$

16. $2\dfrac{1}{4} \times \left(7\dfrac{2}{9} + 5\dfrac{1}{6} - 11\dfrac{5}{8}\right) \times \dfrac{2}{5}$

17. If a watch gains $\frac{3}{7}$ of a minute each hour, how fast will it be in $2\frac{1}{3}$ hours?

18. A car consumes $\frac{2}{33}$ of a gallon of gasoline per mile. How many gallons will it consume in $1,442\frac{1}{10}$ miles?

19. A man owes $\frac{3}{4}$ of $63. If he pays $\frac{3}{7}$ of what he owes, how much will he still owe?

20. If $\frac{1}{8}$ of a cargo worth $7,536 and insured for $\frac{2}{3}$ of its value was damaged while being transported, how much would the insurance company be obliged to pay the shipper?

6.12. Division of Fractions. Division is a special case of subtraction in which the same number is successively subtracted from a given number in order to ascertain the number of times it is contained in the given number. Therefore, to divide 4 by $\frac{2}{5}$ is to determine the number of times that 4 contains $\frac{2}{5}$. There are 5 fifths in one unit. Hence, 2 fifths is contained in one unit as many times as 2 is contained in 5; that is, $\frac{5}{2}$ times. Since $\frac{2}{5}$ is contained $\frac{5}{2}$ times in one unit, it will be contained 4 times $\frac{5}{2}$ in 4 units, or $4 \times \dfrac{5}{2} = \dfrac{4 \times 5}{2} = \dfrac{20}{2} = 10$. Note that if the divisor $\frac{2}{5}$ is inverted and if the inverted divisor $\frac{5}{2}$ is multiplied by the dividend 4, the quotient, $\frac{20}{2} = 10$, is obtained. Thus,

$$4 \div \dfrac{2}{5} = 4 \times \dfrac{5}{2} = \dfrac{4 \times 5}{2} = \dfrac{20}{2} = 10.$$

Similarly, the quotient of $\frac{2}{5} \div 4$ can be obtained in the following manner. One unit is $\frac{1}{4}$ of 4 units; that is, 4 is contained $\frac{1}{4}$ times in one unit. Since 4 is contained $\frac{1}{4}$ times in one unit, it will be con-

tained $\frac{2}{5}$ times $\frac{1}{4}$ in $\frac{2}{5}$ of a unit. Thus $\frac{2}{5} \times \frac{1}{4} = \frac{2 \times 1}{5 \times 4} = \frac{2}{20} = \frac{1}{10}$.
But a whole number may be considered as a fraction whose denominator is 1; that is, $\frac{2}{5} \div 4$ may be written $\frac{2}{5} \div \frac{4}{1}$. Hence, if, as in the preceding case, the divisor $\frac{4}{1}$ is inverted and the inverted divisor $\frac{1}{4}$ is multiplied by the dividend $\frac{2}{5}$, the quotient $\frac{2}{20} = \frac{1}{10}$ is obtained. Thus, $\frac{2}{5} \div 4 = \frac{2}{5} \times \frac{1}{4} = \frac{2 \times 1}{5 \times 4} = \frac{2}{20} = \frac{1}{10}$.

Moreover, to determine the quotient of $\frac{2}{7} \div \frac{3}{5}$ one may reason as follows. There are 5 fifths in one unit. Hence, 3 fifths are contained in one unit as many times as 3 is contained in 5, or $\frac{5}{3}$ times. Since $\frac{3}{5}$ is contained $\frac{5}{3}$ times in one unit, it will be contained $\frac{2}{7}$ times $\frac{5}{3}$ in $\frac{2}{7}$ of a unit. Thus, $\frac{2}{7} \times \frac{5}{3} = \frac{2 \times 5}{7 \times 3} = \frac{10}{21}$. Hence, as before, if the divisor is inverted and then multiplied by the dividend, the quotient is obtained. Thus, $\frac{2}{7} \div \frac{3}{5} = \frac{2}{7} \times \frac{5}{3} = \frac{2 \times 5}{7 \times 3} = \frac{10}{21}$.
Therefore,

To divide a whole number by a fraction, or a fraction by a whole number, or a fraction by another fraction, invert the divisor and multiply.

Example. Find the quotient of: (a) $2 \div \frac{4}{5}$; (b) $\frac{3}{10} \div 9$.

Solution. (a) $2 \div \frac{4}{5} = 2 \times \frac{5}{4} = \frac{\overset{1}{2} \times 5}{\underset{2}{4}} = \frac{5}{2} = 2\frac{1}{2}$. *Ans.*

(b) $\frac{3}{10} \div 9 = \frac{\overset{1}{3}}{10} \times \frac{1}{\underset{3}{9}} = \frac{1}{30}$. *Ans.*

Example. Find the quotient of: (a) $\frac{5}{12} \div \frac{5}{9}$; (b) $8\frac{1}{4} \div 1\frac{1}{8}$.

Solution. (a) $\frac{5}{12} \div \frac{5}{9} = \frac{\overset{1}{5}}{\underset{4}{12}} \times \frac{\overset{3}{9}}{\underset{1}{5}} = \frac{3}{4}$. *Ans.*

(b) $8\frac{1}{4} \div 1\frac{1}{8} = \frac{33}{4} \div \frac{9}{8} = \frac{\overset{11}{33}}{\underset{1}{4}} \times \frac{\overset{2}{8}}{\underset{3}{9}} = \frac{22}{3} = 7\frac{1}{3}$. *Ans.*

Example. A man does $\frac{2}{5}$ of a job in $\frac{3}{4}$ of an hour. What part of the job will he do in 1 hour?

Solution. The man works 3 quarters of an hour; so that, in 1 quarter of an hour he will do $\frac{1}{3}$ of what he does in $\frac{3}{4}$ of an hour. Since in $\frac{1}{4}$ of an hour he does $\frac{1}{3}$ of what he does in $\frac{3}{4}$ of an hour, in 1 hour he will do 4 times $\frac{1}{3}$; that is, $4 \times \frac{1}{3} = \frac{4}{3}$ of what he does in $\frac{3}{4}$ of an hour. But he does $\frac{2}{5}$ of the job in $\frac{3}{4}$ of an hour. Therefore, he will do $\frac{4}{3}$ of $\frac{2}{5}$; that is, $\frac{4}{3} \times \frac{2}{5} = \frac{8}{15}$ of the job in 1 hour;

or

$$\frac{2}{5} \div \frac{3}{4} = \frac{2}{5} \times \frac{4}{3} = \frac{8}{15}. \quad Ans.$$

Example. A tank can be filled by one pipe in 6 hours and by another pipe in $7\frac{1}{3}$ hours. It can be emptied by a waste pipe in 5 hours. If the tank is empty and the three pipes are opened, how long will it take to fill the tank?

Solution. The first pipe will fill $\frac{1}{6}$ of the tank in 1 hour. Since $7\frac{1}{3} = \frac{22}{3}$, the second pipe will fill $\frac{1}{22/3} = \frac{3}{22}$ of the tank in 1 hour. The waste pipe will empty $\frac{1}{5}$ of the tank in 1 hour. Therefore, the three pipes together

will fill $\dfrac{1}{6} + \dfrac{3}{22} - \dfrac{1}{5} = \dfrac{55 + 45 - 66}{330} = \dfrac{34}{330} = \dfrac{17}{165}$ of the tank in 1 hour.

To fill $\frac{17}{165}$ of the tank, the three pipes take 1 hour. To fill $\frac{1}{165}$ of the tank, it will take $\frac{1}{17}$ of an hour and to fill the whole tank or $\frac{165}{165}$, it will take $165 \times \frac{1}{17}$ or $\frac{165}{17}$ hours, that is, $9\frac{12}{17}$ hours. *Ans.*

Example. Milk containing $\frac{1}{25}$ of butter fat is to be mixed with cream containing $\frac{7}{25}$ of butter fat. How many gallons of each must be mixed to make 24 gallons containing $\frac{9}{50}$ of butter fat?

Solution. $\dfrac{9}{50} - \dfrac{1}{25} = \dfrac{9-2}{50} = \dfrac{7}{50}$ = difference of butter fat content

between final mixture and milk. $\dfrac{7}{25} - \dfrac{9}{50} = \dfrac{14-9}{50} = \dfrac{5}{50}$ = difference

of butter fat content between cream and final mixture. Thus for 7 parts of cream there should be 5 parts of milk. Hence, $\frac{7}{12}$ of the total number of gallons should be cream and $\frac{5}{12}$ of the total number of gallons should be milk. Since the mixture is to contain 24 gallons, then $\frac{7}{12} \times 24 = 14$ gallons of cream, and $\frac{5}{12} \times 24 = 10$ gallons of milk. *Ans.*

EXERCISE 28

Perform the indicated operations.

1. $\dfrac{2}{5} \div 6$ **2.** $\dfrac{12}{7} \div 4$ **3.** $8 \div 3\dfrac{1}{5}$

4. $12\dfrac{9}{16} \div 3$ **5.** $\dfrac{12}{25} \div \dfrac{4}{5}$ **6.** $\dfrac{3}{8} \div \dfrac{3}{2}$

7. $1\frac{1}{3} \div 5\frac{3}{4}$ **8.** $7\frac{1}{2} \div 2\frac{6}{7}$ **9.** $6\frac{2}{9} \div 5\frac{1}{4}$

10. $\left(\frac{5}{8} + \frac{3}{4}\right) \div \frac{33}{16}$ **11.** $\frac{17}{24} \div \left(\frac{3}{4} + \frac{2}{3}\right)$ **12.** $\left(\frac{3}{4} - \frac{2}{3}\right) \div \frac{19}{24}$

13. $\left(\frac{4}{5} - \frac{3}{4} + \frac{7}{20}\right) \div \left(2 - \frac{3}{5}\right)$

14. $\left(\frac{2}{5} - \frac{1}{8} - \frac{1}{6}\right) \div \left(\frac{2}{13} + \frac{4}{11}\right)$

15. $\left(\frac{2}{3} - \frac{1}{4}\right) \div \left(4\frac{1}{5} \times \frac{5}{42} \times \frac{1}{6}\right)$

16. $\left(\frac{1}{2} \times \frac{4}{3}\right) \div \left(2\frac{1}{3} - 1\frac{1}{6}\right) \times \left(\frac{1}{2} \div 6\right)$

17. A pumpkin weighs $\frac{9}{10}$ of its weight and $\frac{9}{10}$ of a pound. How many pounds does it weigh?

18. A pipe can fill a tank in 5 hours. Another pipe can fill the same tank in 8 hours. How long will it take both pipes operating together to fill the tank?

19. Allen can do a piece of work in 6 days, working 8 hours a day, and Parker can do the same piece of work in 8 days, working 10 hours a day. How long will it take both men working together to finish the job?

20. If a number is divided by one more than itself, the quotient is $\frac{1}{4}$. If another number is divided by one more than itself, the quotient is one fourth of itself. Find their product.

21. In 4 years a man accumulated a fortune of $64,000. If his fortune increased by $\frac{1}{3}$ each year, how much did he have at the beginning?

22. Anderson & Co., confectioners, wish to mix candy that sells for 35 cents a pound with candy that sells for 75 cents a pound, in order to sell the mixture for 50 cents a pound. If they make up 100 two-pound boxes and 150 one-pound boxes, how many pounds of each candy should be used?

23. Mr. White invested $10,000 in stocks and bonds. He gained $\frac{1}{8}$ of the amount invested in bonds, and lost $\frac{1}{5}$ of the amount invested in stocks. If his net gain was $600, (a) how much money was invested in bonds? (b) how much in stocks?

24. A truck travels at the rate of 20 miles per hour in going to a place and 30 miles per hour in returning. Find the average rate.

25. A businessman looks at his watch on leaving the office for lunch. When he returns, he finds that the hour hand and the minute hand have just changed places from the position they had when he left the office. Find the time (a) when he left; (b) when he came back.

Decimal Fractions

7.1. Decimal Fractions. A common fraction was defined (Section 6.2) as one or more of the equal, integral parts into which a unit may be divided. If the number of equal, integral parts into which the unit may be divided is restricted to 10 or any multiple of 10 by itself, that is, 10, 100, 1,000, and so on, the fraction is called a *decimal fraction*. A decimal fraction is a fraction whose denominator is 10 or a multiple of 10 by itself. Thus, $\frac{7}{10}$, $\frac{3}{100}$, and $\frac{49}{1,000}$ are decimal fractions.

7.2. Notation and Numeration. Place value is used to denote decimal fractions. This principle states that the value of a digit is affected by the place, or position, which it occupies in the numeral. Thus, besides its intrinsic value, every digit has a relative value determined by its place in the numeral (see page 8). Reading from right to left, the digit written in the first place at the right represents units; the digit written immediately to its left has a value which is ten times greater than its value in the unit's place and hence, denotes tens; the digit in the third place indicates hundreds; and so on. This principle is now extended to the writing of decimal fractions by reasoning that if a digit written to the left of the units place has a value which is ten times its value in the units place, then a digit written immediately to the right of the units place must have a value which is one-tenth its value in the units place. In other words, beginning at the units place and reading from left to right, the digit written in the first place immediately to the right of the units place represents tenths, the digit in the second place denotes hundredths, the digit in the third place indicates thousandths, and so on. In order to distinguish the decimal fraction from the integral part of a number, a point, called a *decimal point*, is placed immediately to the right of the units place. The following table illustrates the names of the places from their position relative to the decimal point.

	Hundred Thousands	Ten Thousands	Thousands	Hundreds	Tens	Units	Decimal Point	Tenths	Hundredths	Thousandths	Ten Thousandths	Hundred Thousandths	Millionths
(a)							.	7					
(b)							.		3				
(c)							.		4	9			
(d)					2	3	.	5		6			
(e)				3			.				4		8

Fig. 22.

Thus (a) of Fig. 22 is written .7 or 0.7; (b) is written 0.03; (c) is written 0.049; (d) is written 23.506; and (e) is written 300.000408. Consequently, as shown in (c), 0.049 means 4 hundredths and 9 thousandths. But since $\frac{4}{100} = \frac{40}{1,000}$ and $\frac{40}{1,000} + \frac{9}{1,000} = \frac{49}{1,000}$, the expression 0.049 is usually read 49 thousandths. Similarly, as shown in (d), 23.506 means 2 tens and 3 units (or 23 units), 5 tenths, 0 hundredths, and 6 thousandths and is read 23 and 506 thousandths. Note that the number of decimal places corresponds to the number of zeros in the denominator. Thus, $\frac{49}{1,000} = 0.049$ and $\frac{408}{1,000,000} = 0.000408$. Conversely, $0.203 = \frac{203}{1,000}$ and $0.020437 = \frac{20,437}{1,000,000}$.

A decimal fraction expressed by the use of the decimal point is simply called a *decimal* to distinguish it from the general meaning of the term *decimal fraction*. For example, $\frac{49}{1,000}$ is a decimal fraction, but not a decimal; whereas 0.049 is both a decimal fraction and a decimal.

An expression consisting of a whole number and a decimal is called a *mixed decimal*. Thus, 23.506 is a mixed decimal.

The expression $0.16\frac{2}{3}$ is read sixteen and two-thirds hundredths; $40.5\frac{1}{7}$ is read forty and five and one-seventh tenths. Note that the fraction on the right does not denote a place of its own but always belongs to the place occupied by the digit immediately to the left of it. Difficulty arises when two numbers such as $0.05\frac{1}{2}$ and $5.00\frac{1}{2}$ occur. The number $0.05\frac{1}{2}$ is read five and one-half hundredths and the number $5.00\frac{1}{2}$ should be read five and one-half of a hundredth, or more explicitly, five units and one-half of a hundredth. This difficulty may be avoided by eliminating the fraction whenever possible. In the decimal $0.05\frac{1}{2}$ that is, five and one-half hundredths, the fraction $\frac{1}{2}$ denotes one-half of a hundredth. But $\frac{1}{2} = \frac{5}{10} = .5$, and .5 of a hundredth is by definition equal to $\frac{5}{100}$ which is equal to $\frac{5}{1,000}$. That is, $\frac{1}{2}$ of a hundredth is equal to 5 thousandths. Hence, $0.05\frac{1}{2} = 0.055$, which may be read fifty-five thousandths. Similarly, in the decimal $0.8\frac{1}{4}$, the fraction $\frac{1}{4}$ denotes $\frac{1}{4}$ of a tenth. But $\frac{1}{4} = \frac{25}{100}$ hence, $0.8\frac{1}{4} = 0.825$, read 825 thousandths. Note that if we have a repeating or circulating decimal, the fractional part cannot be eliminated. Thus, $0.16\frac{2}{3} = 0.166\frac{2}{3} = 0.1666\frac{2}{3}$ and so on.

In reading and writing decimals, the following practical rules may be used.

1) *Read the decimal as a whole number and annex the name of the right-hand decimal place.*

Example.   Read: (a) 0.035;   (b) $0.4\frac{1}{5}$.
Solution.
(a) In the decimal 0.035 the name of the right-hand decimal place is thousandths; hence, it is read thirty-five thousandths. *Ans.*
(b) In the decimal $0.4\frac{1}{5}$ the name of the right-hand decimal place is tenths; hence, it is read four and one-fifth tenths. *Ans.*
Or since $0.4\frac{1}{5} = 0.42$, it may also be read forty-two thousandths. *Ans.*

2) *Write the decimal as an integer (or a common fraction). Beginning at the right, with tenths, read to the left, filling vacant places with zeros until the required denomination is reached. Prefix the decimal point to the result.*

Example.   Write as a decimal: (a) nine hundredths;   (b) eight hundred five ten thousandths;   (c) one-half of a hundredth.
Solution.
(a) Write 9. Begin with tenths and read to the left, tenths (9), hundredths (0). Prefix the decimal point. The result is .09 or 0.09. *Ans.*
(b) Write 805. Begin with tenths and read to the left thus, tenths (5), hundredths (0), thousandths (8), ten thousandths (0). Prefix the decimal point. The result is .0805 or 0.0805. *Ans.*
(c) Write $\frac{1}{2}$. Since $\frac{1}{2}$ belongs to the place occupied by the digit immediately to the left of it, begin with tenths (0); hundredths $(0\frac{1}{2})$. Prefix the decimal point. The result is $.00\frac{1}{2}$ or $0.00\frac{1}{2}$. *Ans.*

7.3. Properties of Decimals.   Multiplying or dividing both terms of a common fraction by the same number does not change the value of the fraction (Section 6.6). Thus, if both terms of the decimal fraction $\frac{46}{100}$ are multiplied by 10, the result $\frac{460}{1,000} = \frac{46}{100}$, that is, 0.460 = 0.46. Conversely, if both terms of the decimal fraction $\frac{300}{10,000}$ are divided by 100, the result $\frac{3}{100} = \frac{300}{10,000}$. Hence, 0.0300 = 0.03. It follows that

1) *A decimal is not changed by adding or dropping zeros to the right of its last digit (not zero).*

If in the decimal 0.035 the decimal point is moved one place to the right, the result 0.35 is 10 times greater than 0.035. If in the decimal 0.057 the decimal point is moved two places to the right, the result 5.7 is 100 times greater than 0.057. That is, if in a given decimal the decimal point is moved one place to the right, the relative value of each digit in the result is 10 times greater than its value in the given decimal; if moved two places to the right, 100 times greater; and so on. Hence,

2) *Moving the decimal point one or more places to the right multiplies the decimal by 10 or a multiple of 10 by itself.*

Conversely, if in a given decimal the decimal point is moved one place to the left, the relative value of each digit in the result is $\frac{1}{10}$ of its value in the given decimal; if moved two places to the left, $\frac{1}{100}$; and so on. For example, if in the decimal 0.75 the decimal point is moved one place to the left, the result is 0.075. In the given decimal, the relative value of 7 is tenths and that of 5 is hundredths; in the result, 0.075, the relative value of 7 is hundredths ($\frac{1}{10}$ of $\frac{1}{10}$), and that of 5 is thousandths ($\frac{1}{10}$ of $\frac{1}{100}$). But 0.75 = $\frac{75}{100}$ and 0.075 = $\frac{75}{1,000} = \frac{75}{100} \times \frac{1}{10} = \frac{75}{100} \div 10$. Hence, moving the decimal point one place to the left divides the decimal by 10. Similarly, 4.6 = $\frac{46}{10}$ and 0.046 = $\frac{46}{1,000} = \frac{46}{10} \times \frac{1}{100} = \frac{46}{10} \div 100$ so that moving the decimal point two places to the left divides the decimal by 100. Hence,

3) *Moving the decimal point one or more places to the left divides the decimal by 10 or a multiple of 10 by itself.*

EXERCISE 29

Perform the indicated operations.

1. 0.7 × 10
2. 0.032 × 10
3. 2.034 × 10
4. 0.02 × 100
5. 0.7 ÷ 10
6. 23.01 ÷ 10
7. 32.15 ÷ 100
8. 0.34 × 1,000
9. 0.00157 × 100
10. 72.305 ÷ 1,000
11. 0.268201 × 10,000
12. 1,053.12 ÷ 10,000

Find the missing number in each of the following.

13. $0.05 \times ? = 0.5$ **17.** $0.407 \times ? = 40.7$
14. $31.119 \div ? = 3.1119$ **18.** $962.147 \div ? = 0.0962147$
15. $4.82 \div ? = 0.0482$ **19.** $0.00209 \times ? = 209$
16. $0.1366 \times ? = 13.66$ **20.** $588 \div ? = 0.000588$

7.4. Addition and Subtraction of Decimals.   Recalling that decimals are simply common fractions with a denominator of 10 or a multiple of 10 by itself and that a decimal is not changed by adding or dropping zeros to the right of its last digit, it is easy to see that any number of decimals may be expressed with the same number of decimal places. Thus,

$$2.432 + 0.35 - 0.2037 = 2.4320 + 0.3500 - 0.2037$$

$$= \frac{24{,}320}{10{,}000} + \frac{3{,}500}{10{,}000} - \frac{2{,}037}{10{,}000}$$

$$= \frac{24{,}320 + 3{,}500 - 2{,}037}{10{,}000}$$

$$= \frac{25{,}783}{10{,}000} = 2.5783.$$

Note that the sum can be found directly by reasoning that only like things can be added or subtracted. Hence, if as shown at the left, the decimal points are placed one under the other so that the units will be in one column, the tenths in another column, the hundredths in still another column, and so on filling the vacant places with zeros if desired, we can then proceed as in the addition or subtraction of whole numbers. Note the decimal point of the result is put directly under the other decimal points.

$$
\begin{array}{r}
2.432 \\
+0.350 \\
\hline
2.7820 \\
-0.2037 \\
\hline
2.5783
\end{array}
$$

Example.   Perform the indicated operations: $2.73 + 0.314 - 0.9652$.
Solution.

$$
\begin{array}{r}
2.730 \\
+0.314 \\
\hline
3.0440 \\
-0.9652 \\
\hline
2.0788. \quad Ans.
\end{array}
$$

EXERCISE 30

Perform the indicated operations.

1. $0.7 + 0.21$ **2.** $1.503 + 0.0902$
3. $2.9701 + 0.000356$ **4.** $0.649 - 0.368$

5. $4.85 - 0.0078$ **6.** $0.38 + 3.18 - 1.403$
7. $0.9435 + 0.148 - 0.53467$
                            8. $3.5876 - 0.32 - 1.0503$
9. $5.397 - 1.1184 - 1.91237$
                            10. $8.3107 + 1.975 - 6.13045$
11. $3.51 + 8.932 + 3.1786 - 10.50783$
                            12. $9.14782 - 7.113 + 3.250073 - 1.730006$
13. $8.005833 + 1.01170069 - 3.18373557$
                            14. $12.503 - 10.15007 + 3.1074 - 1.170045$
15. $23.2905 + 0.0067 - 5.29307 - 8.360704$
                            16. $15.063 - 10.5909 - 8.06079 + 7.097015$

17. Henry has \$15.63; Joe has \$3.85 more than Henry; and John, \$4.07 more than Joe. How much do they have altogether?

18. A man had \$145.75. He collected \$61.98 and \$58.08. He then paid a bill for \$97.38. How much did he have left?

19. The fat content of a recipe for cookies is: milk, 0.00167 gm.; eggs, 0.275 gm.; bananas, 0.01 gm.; oats, 0.1498 gm.; and flour, 0.0367 gm. What is the fat content of the cookies?

20. An automobile whose original cost was \$2,503.16 depreciates at the rate of 30 per cent per year. Its trade-in value at the end of four years   is $2503.16 - \$3,003.792 + \$1,351.7064 - \$270.34128 + \$20.275596$. Find its trade-in value.

7.5. Multiplication of Decimals.   As in the case of addition and subtraction, the rule for the multiplication of decimals can be derived by writing the decimals as decimal fractions and applying the rule for the multiplication of common fractions. For example,

$$0.5 \times 0.37 = \frac{5}{10} \times \frac{37}{100} = \frac{5 \times 37}{10 \times 100}$$

$$= \frac{185}{1,000} = 0.185.$$

The denominator of the product, in this case 1,000, must have as many zeros as there are decimal places in the multiplier and the multiplicand together. Since the number of decimal places corresponds to the number of zeros in the denominator, the number of decimal places in the product is equal to the sum of the decimal places in the multiplier and the multiplicand. Therefore,

To multiply decimals, find the product as in the case of whole numbers. Place the decimal point so that the number of decimal places in the product is equal to the sum of the decimal places in the multiplier and the multiplicand.

Example.   Find the product: (a) 17.89×1.25;   (b) 0.138×0.3506.

Solution.   (a)   17.89

             $\underline{1.25}$

             8945

            3578

          $\underline{1789}$

         22.3625.   *Ans.*

The position of the decimal point may also be found by reasoning as follows: 17.89 is between 17 and 18 and 1.25 is between 1 and 2. Since $17 \times 1 = 17$ and $18 \times 2 = 36$, the answer must be between 17 and 36.

         (b)   0.3506

             $\underline{0.138}$

            28048

          10518

       $\underline{3506}$

     0.0483828.   *Ans.*

Note: Since the product 483828 contained fewer figures than the required number of decimal places, we prefixed one zero so that the product would have the necessary seven decimal places. Again, the position of the decimal point could be obtained by reasoning that 0.3506 is between 0.3 and 0.4, and 0.138 is between 0.1 and 0.2. Since $0.3 \times 0.1 = 0.03$ and $0.4 \times 0.2 = 0.08$, the answer must be between 0.03 and 0.08.

EXERCISE 31

Perform the indicated operations.

1. 0.7×0.4
2. 0.85×0.14
3. 4.38×53.41
4. 19.87×0.1364
5. 0.0058×5.4581

6. 413.007×0.000512
7. $(0.879 - 0.0021) \times 1.0905$
8. $(5.38 + 3.005 + 0.07) \times 0.06$
9. $(142.1 + 0.01 + 0.0001) \times 8.01$
10. $(5.25 + 0.0789 - 0.03) \times 8.2004$

11. Eighteen metal slitting saws, 0.1875 of an inch thick, are to be packed in a box. If one is placed on top of another, what must the depth of the box be?

12. Copper wire weighs 39.68 pounds per 1,000 feet. What will the weight of a line from Albany to New York be if the distance is 147 miles? (1 mile = 5,280 feet)

13. The distance from New Orleans to Los Angeles is 1,938 miles. A car consumes 0.0715 of a gallon of gasoline per mile. (a) How many gallons of gasoline will be consumed on this trip? (b) If a gallon of gasoline costs $0.3725, how much will the gasoline used for this trip cost?

14. The compound amount, that is, the total sum of money accumulated on a principal of $1.00 invested at $\frac{1}{2}$ per cent per month compounded monthly for 8 years, is $1.61414271. Find the compound amount of $307.14 invested at the same rate and for the same period of time.

15. A machine whose original cost was $1.00 depreciated at the annual rate of 20 per cent. At the end of 4 years, its scrap value was $0.4096. What is the scrap value, at the end of 4 years, of a machine whose original cost was $1,587.34, if it depreciates at the same rate?

7.6. Division of Decimals. To derive the rule for the division of a decimal by an integer, we express the decimal as a common fraction. Thus,

$$0.8717 \div 23 = \frac{8,717}{10,000} \div 23 = \frac{8,717}{10,000} \times \frac{1}{23} = \frac{8,717}{23} \times \frac{1}{10,000}$$

$$= 379 \times \frac{1}{10,000} = \frac{379}{10,000} = 0.0379.$$

Note that the number of *decimal places in the quotient is the same as the number of decimal places in the dividend.* Moreover, if we divide as shown at the left, the decimal point in the quotient is placed directly above the decimal point in the dividend.

```
      0.0379
23)0.8717
     69
    ___
    181
    161
    ___
    207
    207
    ___
```

Frequently, the quotient cannot be expressed without a remainder. In that case, the division is carried out to the desired number of decimal places by *rounding off* the quotient, that is, by dropping all decimals after a certain, significant place. If the first digit dropped in the quotient is less than 5, the preceding digit is not changed. If the first digit dropped is greater than or equal to 5 and the succeeding digit is not zero, the preceding digit is increased by 1. If the first digit dropped is equal to 5 and all succeeding digits are zero, the preceding digit is unchanged, or increased by 1, as is necessary to leave the preceding digit even.

Example. Perform the indicated operation to the nearest ten thousandth: (a) 0.4505 ÷ 17; (b) 450.5679 ÷ 54.

Solution. (a)
```
      0.0265   Ans.
17)0.4505
    34
    ___
    110
    102
    ___
     85
     85
```

(b) $450.5679 \div 54$

$$
\begin{array}{r}
8.34385 \quad Ans. \\
54\overline{)450.5679} \\
\underline{432} \\
185 \\
\underline{162} \\
236 \\
\underline{216} \\
207 \\
\underline{162} \\
459 \\
\underline{432} \\
270 \\
\underline{270}
\end{array}
$$

If the divisor is not an integer, the problem can be expressed in the form of common fractions and then reduced to the case of division of a decimal by an integer. Thus,

$$0.084 \div 0.7 = \frac{84}{1,000} \div \frac{7}{10} = \frac{84}{1,000} \times \frac{10}{7} = \frac{84}{100} \times \frac{1}{7}$$

$$= 0.84 \div 7.$$

Similarly,

$$3.7 \div 0.08 = \frac{37}{10} \div \frac{8}{100} = \frac{37}{10} \times \frac{100}{8} = 37 \times \frac{10}{8}$$

$$= \frac{370}{8} = 370 \div 8.$$

Note that if both the dividend and the divisor are multiplied by 10 or 100, respectively, the operation remains unchanged and the divisor becomes an integer. In general, if both the dividend and the divisor are multiplied by 10 or by the least multiple of 10 by itself that will make the divisor an integer, the operation is reduced to the division of a decimal by an integer, and we can proceed as before.

Example. Perform the indicated operation to the nearest thousandth: (a) $0.0833 \div 0.7$; (b) $3.1683 \div 0.15$; (c) $56 \div 0.114$.

Solution. (a) $0.0833 \div 0.7 = 0.833 \div 7$

$$
\begin{array}{r}
0.119 \quad Ans. \\
7\overline{)0.833}
\end{array}
$$

(b) $3.1683 \div 0.15 = 316.83 \div 15$

$$\begin{array}{r} 21.122. \quad Ans. \\ 15\overline{)316.83} \\ \underline{30} \\ 16 \\ \underline{15} \\ 18 \\ \underline{15} \\ 33 \\ \underline{30} \\ 30 \\ \underline{30} \end{array}$$

(c) $56 \div 0.114 = 56000 \div 114$

$$\begin{array}{r} 491.228. \quad Ans. \\ 114\overline{)56000} \\ \underline{456} \\ 1040 \\ \underline{1026} \\ 140 \\ \underline{114} \\ 260 \\ \underline{228} \\ 320 \\ \underline{228} \\ 920 \\ \underline{912} \\ 80 \end{array}$$

Example. A certain dress material costs $2.98 per square yard before it is pre-shrunk. How much will it cost after shrinking, if it loses 0.035 of a square yard?

Solution. Since 0.035 of a square yard is shrinkage, the remainder, after 0.035 of a square yard has been deducted, is 0.965 of a square yard. If 0.965 of a square yard cost $2.98, a square yard will cost $2.98 \div 0.965 =$ $3.09. *Ans.*

EXERCISE 32

Perform the indicated operation to the nearest thousandth.

1. $0.064 \div 4$
2. $0.81 \div 27$
3. $97.52 \div 4$
4. $281.23 \div 156$
5. $561.09 \div 256$
6. $125.9035 \div 19$
7. $50.1208 \div 143$
8. $0.4842 \div 0.9$
9. $0.20532 \div 0.87$
10. $1.41276 \div 0.305$
11. $1.274534 \div 0.124$
12. $39.18033 \div 3.15$
13. $82 \div 0.035$
14. $132 \div 25.06$

15. Square bars of steel 0.75 of an inch thick weigh 1.913 pounds per foot. (a) How many bars can be packed, one on top of another, in a box 6 inches deep? (b) If each bar is 2 feet long, what will the total weight of the bars in the box be?

16. If a car consumes 0.0475 of a gallon of gasoline per mile, how many miles does the car travel when it consumes 66.5 gallons of gasoline?

17. Water flows from a pipe at the rate of 5.7167 gallons per minute. How long will it take this pipe to fill a tank containing 50.25 gallons?

18. A man does 0.1875 of a job in 0.75 of an hour. What part of the job will he do in 1 hour?

19. Water flows from a pipe at the rate of 2.2204 gallons per minute and from another pipe at the rate of 1.6526 gallons per minute. If the tank is empty and both pipes are opened, how long will it take to fill a tank with a capacity of 112.388 gallons?

20. A set of dinnerware costs $18.53. If a loss of $0.10 per dollar is estimated due to breakage in handling, find the actual cost per set after handling.

7.7. Reduction of Decimals. Any process which changes a fraction from one form to another, without changing its value, is called reduction (Section 6.7). The two most important ways to reduce decimals are: (1) changing a decimal to a common fraction; and (2) changing a common fraction to a decimal.

7.7–1. Changing a Decimal to a Common Fraction. Consider the decimal 0.85. This decimal is read 85 hundredths and hence may be written $\frac{85}{100}$. Reducing this fraction to its lowest terms, we obtain $\frac{85}{100} = \frac{17}{20}$ so that $0.85 = \frac{17}{20}$. Therefore,

To change a decimal to a common fraction express the decimal as a decimal fraction, that is, as a common fraction with a denominator of 10 or a multiple of 10 by itself, and then reduce it to its lowest terms.

Example. Reduce each of the following decimals to a common fraction: (a) 0.75;   (b) 5.875;   (c) $0.12\frac{1}{4}$.

Solution.   (a) $0.75 = \dfrac{75}{100} = \dfrac{75 \div 25}{100 \div 25} = \dfrac{3}{4}$. *Ans.*

(b) $5.875 = 5\dfrac{875}{1,000} = 5\dfrac{7}{8}$. *Ans.*

(c) $0.12\dfrac{1}{4} = \dfrac{12\frac{1}{4}}{100} = \dfrac{\frac{49}{4}}{100} = \dfrac{49}{400}$. *Ans.*

or $0.12\dfrac{1}{4} = 0.1225 = \dfrac{1225}{10,000} = \dfrac{49}{400}$. *Ans.*

7.7–2. Changing a Common Fraction to a Decimal. To change a common fraction to a decimal fraction, reduce the given fraction to an equivalent fraction with a denominator of 10 or a multiple of 10 by itself. For example, to reduce $\frac{3}{4}$ to a decimal fraction with a denominator of 100, use the rule on page 101. Thus, $100 \div 4 = 25$, so that $\dfrac{3}{4} = \dfrac{3 \times 25}{4 \times 25} = \dfrac{75}{100}$, but $\dfrac{75}{100} = 0.75$; hence, $\dfrac{3}{4} = 0.75$. Similarly, to reduce $\frac{2}{3}$ to an equivalent fraction with a denominator of 100, divide 100 by 3, thus $100 \div 3 = 33\frac{1}{3}$ so that $\frac{2}{3} = \frac{2 \times 33\frac{1}{3}}{3 \times 33\frac{1}{3}} = \frac{66\frac{2}{3}}{100}$. But $\frac{66\frac{2}{3}}{100} = 0.66\frac{2}{3}$; hence, $\frac{2}{3} = 0.66\frac{2}{3}$.

Note that in the division shown below, $3 \div 4 = 0.75$ and $2 \div 3 = 0.66\frac{2}{3}$. In other words, if we divide the numerator by the denominator, the decimal is obtained directly. Therefore,

To change a common fraction to a decimal, divide the numerator by the denominator.

$$
\begin{array}{r}
0.75 \\
4)\overline{3.00} \\
\underline{28} \\
20 \\
\underline{20} \\
\end{array}
\qquad\qquad
\begin{array}{r}
0.66 \\
3)\overline{2.00} \\
\underline{18} \\
20 \\
\underline{18} \\
2 \\
\end{array}
$$

Example.   Reduce each of the following fractions to a decimal: (a) $\frac{3}{20}$; (b) $\frac{21}{8}$;   (c) $\frac{1}{6}$.

Solution.   (a)
$$
\begin{array}{r}
0.15 \\
20)\overline{3.00} \\
\underline{20} \\
100 \\
\underline{100} \\
\end{array}
$$
(b)
$$
\begin{array}{r}
2.625 \\
8)\overline{21.000} \\
\underline{16} \\
50 \\
\underline{48} \\
20 \\
\underline{16} \\
40 \\
\underline{40} \\
\end{array}
$$
(c)
$$
\begin{array}{r}
0.16 \\
6)\overline{1.00} \\
\underline{6} \\
40 \\
\underline{36} \\
4 \\
\end{array}
$$

0.15.   *Ans.*   2.625.   *Ans.* $0.16\frac{4}{6} = 0.16\frac{2}{3}$.   *Ans.*

Example.   The market prices of stocks are published daily on the financial pages of newspapers. The prices are listed in whole dollars and fractions thereof. Express each of the following market prices of stocks as a decimal: (a) Ajax Petrol $\frac{7}{8}$;   (b) Decca Records, Inc. $17\frac{3}{8}$; (c) Standard Oil Co. (Ind.) $47\frac{3}{4}$.

Solution.   (a)
$$
\begin{array}{r}
0.875 \\
8)\overline{7.000} \\
\underline{64} \\
60 \\
\underline{56} \\
40 \\
\underline{40} \\
\end{array}
$$
(b)
$$
\begin{array}{r}
0.375 \\
8)\overline{3.000} \\
\underline{24} \\
60 \\
\underline{56} \\
40 \\
\underline{40} \\
\end{array}
$$
(c)
$$
\begin{array}{r}
0.75 \\
4)\overline{3.00} \\
\underline{28} \\
20 \\
\underline{20} \\
\end{array}
$$

$0.875.   *Ans.* $17.375.   *Ans.* $47.75.   *Ans.*

7.8. Repeating Decimals. From the preceding discussion, it is evident that no matter how many zeros we annex to the numerator, the denominator is not always contained an exact number of times in the numerator. When the division ends with an exact quotient, the fraction being reduced can be expressed as an exact decimal, but when the division does not end in an exact quotient, a digit or a succession of digits will repeat indefinitely. In fact, if the denominator has a prime factor different from 2 or 5, the quotient cannot be exact. A decimal with a digit or a set of digits that repeats indefinitely is called a *repeating* or *circulating decimal*. The part of the circulating decimal which repeats indefinitely is indicated by placing a dot over the recurring digit, or over each of the digits in the recurring set, or by placing three dots following the decimal. Thus, $\frac{2}{3} = 0.6\frac{2}{3}$ may be written $0.\dot{6}$ or $0.66\cdots$. Similarly, $\frac{2}{7} = 0.285714\frac{2}{7}$ can be written $0.\dot{2}8571\dot{4}$ or

$$0.285714285714\cdots.$$

Operations with repeating decimals can be exact if the common fraction equivalent to the repeating decimal is used. However, in most cases, this degree of accuracy is not necessary, so we round off the figures to obtain an answer which is correct to the necessary number of decimal places (page 124). Thus, if 3 cans of tomato sauce sell for $0.29, the price per can is $0.29 \div 3 =$ $0.09\frac{2}{3}$. The exact value of 4 cans can be obtained by using the common fraction equivalent to $0.09\frac{2}{3}$, that is, $\frac{29}{300}$; thus $4 \times \frac{29}{300} = \frac{29}{75} = \$0.38\frac{2}{3}$. However, since the answer must be rounded off to the nearest cent, that is, to two decimal places, the price per can ($0.09\frac{2}{3}$) may be taken as $0.097 and this number will give an answer which, although not exact, is correct for all practical purposes. Thus, $4 \times 0.097 = 0.388$ or $0.39.

Whenever repeating decimals occur, the value of the common fraction differs from the value of the decimal form by a fractional part of the fractional unit expressed as a decimal. Thus, $\frac{2}{3} = 0.6\frac{2}{3} = 0.66\frac{2}{3} = 0.666\frac{2}{3} = 0.6666\frac{2}{3}$ and so on. Hence,

if $\frac{2}{3} = 0.6\frac{2}{3}$, then $\frac{2}{3}$ differs from 0.6 by $\frac{2}{3}$ of $\frac{1}{10}$ or $\frac{2}{30}$;

if $\frac{2}{3} = 0.66\frac{2}{3}$, then $\frac{2}{3}$ differs from 0.66 by $\frac{2}{3}$ of $\frac{1}{100}$ or $\frac{2}{300}$;

if $\frac{2}{3} = 0.666\frac{2}{3}$, then $\frac{2}{3}$ differs from 0.666 by $\frac{2}{3}$ of $\frac{1}{1,000}$ or $\frac{2}{3,000}$;

if $\frac{2}{3} = 0.6666\frac{2}{3}$, then $\frac{2}{3}$ differs from 0.6666 by $\frac{2}{3}$ of $\frac{1}{10,000}$ or $\frac{2}{30,000}$.

Observe that the difference between the value of the fraction $\frac{2}{3}$ and the value of the decimal form decreases as the number of decimal places increases. In fact, by using a large number of decimal places we can make the difference between the value of the fraction and the value of the decimal as small as we please.

Example. Reduce each of the following fractions to (1) a repeating decimal; (2) a decimal to the nearest hundredth. (a) $\frac{1}{12}$. (b) $\frac{3}{7}$. (c) $5\frac{5}{9}$.

Solution. (a)

$$
\begin{array}{r}
0.083 \\
12\overline{)1.000} \\
\underline{96} \\
40 \\
\underline{36} \\
4
\end{array}
$$

(1) 0.08$\dot{3}$. *Ans.*

(2) 0.08. *Ans.*

(b)

$$
\begin{array}{r}
0.428571 \\
7\overline{)3.000000} \\
\underline{28} \\
20 \\
\underline{14} \\
60 \\
\underline{56} \\
40 \\
\underline{35} \\
50 \\
\underline{49} \\
10 \\
\underline{7} \\
3
\end{array}
$$

(c)

$$
\begin{array}{r}
0.55 \\
9\overline{)5.00} \\
\underline{45} \\
50 \\
\underline{45} \\
5
\end{array}
$$

(1) 5.$\dot{5}$. *Ans.*
(2) 5.56. *Ans.*

(1) 0.$\dot{4}$285$\dot{7}$1. *Ans.*

(2) 0.43. *Ans.*

EXERCISE 33

Reduce each of the following decimals to a common fraction.

1. 0.2
2. 0.04
3. 0.25
4. 0.45
5. 0.3125
6. 0.375
7. $0.416\frac{2}{3}$
8. 24.075
9. 3.0025
10. 1.025
11. $13.83\frac{1}{3}$
12. $9.714285\frac{5}{7}$

Reduce each of the following fractions to a decimal and round off to the nearest thousandth, where necessary.

13. $\frac{1}{5}$ 14. $\frac{5}{8}$ 15. $\frac{3}{5}$ 16. $\frac{3}{20}$ 17. $\frac{1}{40}$

18. $\dfrac{19}{20}$ **19.** $\dfrac{5}{6}$ **20.** $\dfrac{5}{12}$ **21.** $14\dfrac{2}{7}$ **22.** $\dfrac{7}{11}$

23. $\dfrac{5}{13}$ **24.** $\dfrac{8}{17}$ **25.** $\dfrac{5}{24}$ **26.** $\dfrac{1}{6}$ **27.** $\dfrac{7}{15}$

7.9. Accuracy of Computations. In most business transactions, fractions of a penny are usually ignored, for the result must be rounded off to the nearest cent. However, in science, where measurements are statistical in character, a high degree of accuracy is necessary but often difficult to achieve. For example, the steel tapes used by surveyors are usually 100 feet long and graduated to 0.01 of a foot. Actually, these tapes are said to be 100 feet long at 62°F., with a probable error of about 0.0000065 of a foot per Fahrenheit degree. But corrections for temperature are uncertain, since the temperature of the tape in an open field cannot be determined with any great accuracy. Hence, it is deceptive to state that a certain distance, measured under normal conditions with a surveyor's tape, is *exactly* 46.73 feet in length. What we mean is that, under existing conditions, this distance seems to be nearer to the mark 46.73 than it is to the mark 46.72 or 46.74.

Even in abstract arithmetical computations, exact results are possible only when the numbers dealt with are integers or common fractions. In fact, there are many cases in which the precise answer cannot be obtained no matter how far an indicated operation may be carried out. For example, there is no integer or common fraction whose square is 2; hence, the square root of 2 cannot be exactly obtained as a terminating decimal.* Hence, perfect accuracy is seldom obtainable.

The *significant figures* of a number are those digits which determine its accuracy. For example, if a statistical report states that "Automobiles, trucks, and buses operated a total of 456,000,000,000 miles over U.S. roads and streets," this number is assumed to be accurate to a billion miles, that is, this statement is true, give or take one billion miles. Here the significant figures are 4, 5, and 6. If we are working with numbers accurate to three places, the significant figures of a number such as 450 are 4, 5, and 0; and those of 0.450 are also 4, 5, and 0. Moreover, in 0.45,

*See William H. Durfee, *Fundamentals of College Algebra* (New York: Macmillan, 1960).

the 0 is not significant, but in 0.045, the 0 in the tenths place is significant.

In the addition or subtraction of several numbers, the accuracy of the least precise number will determine the accuracy of the result. For example, if the weight of an object is given to the nearest pound, that is, to the nearest unit, and the weight of another object is given to the nearest tenth of a pound, it is evident that the sum, or difference, of these weights can only be correct to the nearest pound. Thus in the operation 3 pounds + 2.3 pounds = 5.3 pounds, the answer is accurate only to the nearest pound, that is, 5 pounds. This is also true in the multiplication or division of numbers.† In general, if the least accurate of several numbers has n significant figures, the accuracy of the result of any fundamental operation of arithmetic is to n digits. Thus, the operation $137.85 \times 2.6 = 358.410$ is accurate to two significant figures, that is, to 350, for the number 2.6 has only two significant figures. However, if by writing 2.6000 we indicate that this number is accurate to four decimals, the product $137.85 \times 2.6000 = 358.41$ will be accurate to five figures.

The *reciprocal* of a given number is the number whose product with the given number is equal to 1. Hence, the reciprocal of a given number is 1 divided by that number. Thus, the reciprocal of 5 is $\frac{1}{5}$; the reciprocal of $\frac{1}{3}$ is $1 \div \frac{1}{3} = 1 \times \frac{3}{1} = 3$; the reciprocal of $\frac{3}{7}$ is $1 \div \frac{3}{7} = 1 \times \frac{7}{3} = \frac{7}{3}$.

In order to simplify the writing and handling of very small numbers, we define the notation a^{-n} as the reciprocal of the nth power of any given number a, that is,

$$a^{-n} = \frac{1}{a^n}$$

For example, the reciprocal of 4^2 is $\frac{1}{4^2}$ and this can be written 4^{-2}; that is, $4^{-2} = \frac{1}{4^2}$. Moreover,

$$0.0025 = \frac{25}{10,000} = \frac{25}{10^4} = 25 \times 10^{-4}.$$

Scientists can briefly express very large or very small numbers in the form $a10^n$ or $a10^{-n}$, where a represents a number between 1 and 10 and n is any integer. Thus, the number 456,000,000,000

† For a proof of this statement, see pages 162–163.

written in this scientific notation is 4.56×10^{11}, and 0.0025, is 2.5×10^{-3}.

EXERCISE 34

Write the following numbers using scientific notation.

1. 1 kilometer = 0.62137 miles.
2. Velocity of light = 186,330 miles per second.
3. Charge of 1 electron = 0.0000000004774 electrostatic units.
4. 1 mol = 606,000,000,000,000,000,000,000 molecules.
5. 1 mean gram-calorie = 41,834,000 ergs.

Perform the indicated operations and give the accurate answer.

6. 4.7×2.6 **11.** $17.8 \div 0.044$
7. 3.15×7.3 **12.** $(14 \times 32) \div 44$
8. 6.03×0.142 **13.** $(37 \times 1.03) \div 2.3$
9. $8.4 \div 2.3$ **14.** $(18.374 + 4.358) \div 2.273$
10. $60.7 \div 1.25$ **15.** $(34.03 + 0.3952 - 2.21) \times 1.02$
16. $(142.1 + 0.01 + 0.0001) \times 8.01$
17. $(16.7313 \div 3.891) + (0.0073 \times 1.16)$
18. $(5.25 + 0.0789 - 2.03) \times 8.2004$
19. $2.209 - 1.0489 - 5.00357 + 8.0185$
20. $(7.63 \times 12.2 - 5.186) + (10.344 \div 0.15516)$

7.10. Short Methods. The solution of scientific and business problems usually requires that answers be given to a specified number of decimal places. Considerable time and labor can be avoided by omitting figures which do not influence the accuracy of the result.

For example, the product 2.14327×1.43572 obtained by the usual method is shown at the left. If accuracy only to the nearest thousandth is desired, the work shown to the right of the vertical line is unnecessary. These figures may be eliminated as follows. The multiplicand is greater than 2 but less than 3, that is, 2.14327 lies between 2 and 3. The multiplier, 1.43572, lies between 1 and 2. Thus, the product lies between 2×1 and 3×2 or between 2 and 6. Hence, the product has one figure to the left of the decimal point. In order that the answer be accurate to the nearest thousandth, that is, to three decimal places, it is necessary to obtain four decimals in the product. It follows that the product must have one figure to the left of the decimal plus the four decimals needed, or a total of five figures. In the

```
      2.14327
      1.43572
     |4 28654
   15|0 0289
  107|1 635
  642|9 81
 8573|0 8
2 1432|7
3.0771|3 56044
```

short method that follows, the number of figures in the multiplicand fixes the number of figures in the product. Moreover, the order in which the numbers are multiplied is immaterial provided place values are kept in mind. Hence, we choose the first five digits in each of the factors and then write the multiplier in reverse order, thus,

$$21432$$
$$\underline{75341}$$

1) Multiply by 1 as usual, as shown below. Then cross out the 1 in the multiplier and the 2 above it in the multiplicand. Thus,

$$2143\cancel{2}$$
$$\underline{7534\cancel{1}}$$
$$\overline{21432}$$

2) Multiply by 4, taking care to carry over the digit from the multiplication of the figure crossed out in the multiplicand, in the preceding step. Thus, since 2 was crossed out, the product would have been 4 × 2 = 8. Carry 1, for 8 is nearer 10 than 0. Then, as shown below, multiply by 4 as usual. Thus, 4 × 3 = 12 and 1 carried is 13. Write 3 under the 2, carry 1. Then, 4 × 4 = 16 and 1 carried is 17. Write 7 under the 3, carry 1, and so on. Cross out the 4 in the multiplier and the 3 above it in the multiplicand.

$$214\cancel{3}\cancel{2}$$
$$\underline{7534\cancel{1}}$$
$$\overline{21432}$$
$$8573$$

3) Proceed in this manner until the multiplication is completed, as shown below.

$$2\cancel{1}\cancel{4}\cancel{3}\cancel{2}$$
$$\underline{7\cancel{5}\cancel{3}\cancel{4}\cancel{1}}$$
$$\overline{21432}$$
$$8573$$
$$643$$
$$107$$
$$15$$
$$\overline{30770}$$

Since the product lies between 2 and 6, the answer to the nearest thousandth is 3.077.

Taking another example, the quotient of 3.07713 ÷ 1.43572 can be obtained to the nearest thousandth as follows. Since the dividend is between 3 and 4 and the divisor is between 1 and 2, the quotient is between 4 ÷ 2 and 3 ÷ 1, or between 2 and 3. For an answer accurate to the nearest thousandth, we must have one figure to the left of the decimal point plus the four decimals needed, or a total of five figures. In this short method, the number of figures in the quotient is equal to the number of figures in the divisor. Hence, we choose the first five figures of the divisor and as many figures in the dividend as are necessary to begin the division. We then write,

$$14357 \overline{)30771}$$

1) The first digit in the quotient is 2. Write 2 under the 7 in the divisor, as shown below. Then multiply 14357 by 2 and subtract the product, 28714, from 30771, as usual. Cross out the 7 in the divisor above the 2 in the quotient. Thus,

$$
\begin{array}{r}
14357\overline{)30771} \\
2 \; \underline{28714} \\
2057
\end{array}
$$

2) The remaining divisor, 1435, is contained once in the remainder 2057. Write 1 under the 5 in the quotient, as shown below. Multiply 1435 by 1, taking care to carry over the digit from the multiplication of the figure crossed out in the divisor in the preceding step. Thus, since 7 was crossed out, the product would have been $1 \times 7 = 7$. Carry 1, for 7 is nearer 10 than 0. Then multiply as usual. Thus, $1 \times 5 = 5$ and 1 carried is 6. Write 6 under 7 in 2057. Then $1 \times 3 = 3$, and so on. Subtract the product, 1436, from 2057. Cross out the 5 in the divisor above the 1 in the quotient, leaving a new divisor of 143.

$$
\begin{array}{r}
14357\overline{)30771} \\
12 \; \underline{28714} \\
2057 \\
\underline{1436} \\
621
\end{array}
$$

3) Proceed in this manner until the division is completed, as shown at the top of the following page.

$$14357)\overline{30771}$$
$$33412 \quad \underline{28714}$$
$$2057$$
$$\underline{1436}$$
$$621$$
$$\underline{574}$$
$$47$$
$$\underline{43}$$
$$4$$
$$\underline{4}$$

Note that when the last step is reached, the number 3 (not 4) is taken as a divisor, since the carry from the crossed-out number must be taken into account. Thus, 3 times the crossed-out 4 is 12, carry 1; then $3 \times 1 = 3$ and 1 carried is 4. The quotient is under the divisor in reverse order, so that the quotient is 21433. Since it must have one figure to the left of the decimal point, the answer to the nearest thousandth is 2.143.

EXERCISE 35

Perform each of the following operations using the short method.

1. 1.09344×1.81402 to the nearest ten thousandth;
2. 3.94609×1.94790 to the nearest thousandth;
3. 4.53804×2.52695 to the nearest ten thousandth;
4. 1.41852×5.74349 to the nearest hundredth;
5. 5.51602×0.61391 to the nearest hundredth;
6. $3.55567 \div 1.62889$ to the nearest thousandth;
7. $1,583.25 \div 2.90503$ to the nearest hundredth;
8. $2,034.78 \times 1.80611$ to the nearest hundredth;
9. $4,763.82 \div 3.11865$ to the nearest thousandth;
10. $5,743.49 \div 0.55839$ to the nearest hundredth.

7.11. Square Root–General Method. The method of finding the square root of an integer, that is a perfect square, was demonstrated in Section 5.8. The square root of any other number can be found by the same method. First, separate the number into groups as follows. Beginning at the decimal point, separate the number into groups of two figures each. Indicate groups from the decimal point to the left for the integral part of the root, and from the decimal point to the right for the decimal part of the root. Zeros may be added as needed to obtain as many decimal places in the answer as are required. Any single figure remaining at the left of the integral part, after the groups have been indicated, is counted as a group. Each group of the decimal part must contain two figures.

Example. Find $\sqrt{943.834}$ to the nearest hundredth.

Solution. 1) Separate the number into groups of two figures each, as

(a)
```
     3 0.
  3 | 9'43'83'40'00
    | 9
  6 | 43
```

(b)
```
      3 0. 7
   3 | 9'43'83'40'00
     | 9
 607 | 43 83
     | 42 49
     |  1 34
```

shown in (a). Note that 9, at the left in the integral part, is considered as a group. However, a zero is annexed to the 4 in the decimal part to form a group of two figures, and two more zeros are added to obtain the required three decimal places in the root.

2) Find the number whose square is either equal to or less than the first left-hand group, 9. The number is 3 so write 3 above the radicand as the first digit of the root and write it again at the left of the radicand. Square 3, that is, get the product $3 \times 3 = 9$; write it under the 9 in the radicand and subtract, obtaining a remainder of 0. Bring down the next group, 43.

3) Multiply the first digit in the root by 2. Thus, $3 \times 2 = 6$, and write 6 to the left of 43. Find how many times 6 is contained in 43, exclusive of the right-hand digit, that is, how many times 6 is contained in 4. Since 6 is not contained in 4, write 0 as the second digit in the root and place the decimal point after the 0.

4) As shown in (b), bring down the next group, 83, and multiply all the digits in the root by 2. Thus, $30 \times 2 = 60$. Write 60 to the left of 4383. Find how many times 60 goes into 4383, exclusive of the right-hand digit, that is, how many times 60 is contained in 438. The answer is 7. Write 7 as the tenths digit in the root, and write it also to the right of 60, getting 607. Find the product $607 \times 7 = 4249$ and subtract it from 4383, obtaining 134. Proceed as before. The completed operation is shown in (c).

(c)
```
         3 0. 7 2 1
     3 | 9'43'83'40'00
       | 9
   607 | 43 83
       | 42 49
  6142 |  1 34 40
       |  1 22 84
 61441 |    11 56 00
       |     6 14 41
       |     5 41 59
```

Hence, $\sqrt{943.834}$ = 30.72. *Ans.*

EXERCISE 36

Compute the square root of each of the following numbers to the nearest hundredth.

1. 250
2. 69
3. 379
4. 1,220

5. 7,940
6. 13.48
7. 875.467
8. 1,750.4372

9. 0.73261
10. 1,603.5276
11. 28,160.1961
12. 10,926.5209

13. A boy had $0.60. He did a job for his father, who paid him $0.18 less than the $2.25 he asked for. His brother then gave him $0.15 more than the $1.75 he owed him. How much more does he need to get $5.75?

14. A merchant ordered 316.5 yards of material. He received a first consignment of 50.35 yards; a second of 15.14 yards more than the first; a third equal to the total amount he received on the first two; and a fourth equal to the remainder. How many yards did he receive in the fourth consignment?

15. The compound amount, that is the total sum of money accumulated on a principal, of $1.00 invested at $\frac{7}{8}$ per cent per quarter compounded quarterly for 15 years is $1.68660298. Find the compound amount of $716.09 invested at the same rate and for the same period of time.

16. The monthly payments that must be made to repay a debt of $2,368.46 in 2 years, with interest at $\frac{1}{2}$ per cent per month compounded monthly, is equal to 2,368.46 ÷ 22.56286622. Find the value of the monthly payment.

17. A house yields a net rent of $115 a month. If money is worth 5 per cent per year, the value of this house is now 115 ÷ 0.00407412. Find the value of the house.

18. In computing the square root of an integer, the last completed dividend was found to be 1376. Find the integer and the square of this integer.

19. Add one-third of twelve to four-fifths of seven. The result is eleven.

20. If a man goes to get exactly 4 gallons of water with a 5 and a 3 gallon measure, how can he measure it?

Ratio, Proportion, and Percentage

8.1. Ratio. The *ratio* of two similar quantities is the numerical measure of their relation. The relation of 6 to 2 may be shown in two ways: (1) by stating that 6 is *four more* than 2; and (2) by stating that 6 is *three times* 2. In other words, the relation of two similar quantities may be measured by subtraction or by division. The numerical measure obtained by subtraction, 6 − 2 = 4, is called the *arithmetic ratio*, whereas the numerical measure obtained by division, 6 ÷ 2 = 3, is called the *geometric ratio*. Unless otherwise stated, the word *ratio* denotes a geometric ratio. Therefore the ratio of one quantity to another similar quantity is the result of dividing the first quantity by the second. Thus the ratio of 6 to 2 is 6 ÷ 2 = 3, and the ratio of 3 to 5 is 3 ÷ 5 = 0.6. Since a ratio measures the numerical relations of two similar quantities, it is always an abstract number. For example, the ratio of 6 feet to 2 feet is *not* 3 feet, but simply 3.

8.2. Numeration and Notation. The two similar quantities that make up a ratio are called the *terms* of the ratio. The first term is called the *antecedent* and the second term is called the *consequent*. Thus in the ratio of 6 to 2, the numbers 6 and 2 are the terms of the ratio; 6 is the antecedent and 2 is the consequent. The antecedent 6 is compared with the consequent 2; hence, the consequent is the basis of comparison.

Ratios are expressed by the symbol : placed between the two terms being compared or, since a fraction implies division, in the form of a fraction. For example, the ratio of 6 to 2 may be written 6:2 or $\frac{6}{2}$. Since, by definition, a ratio can be written in the form of a fraction, ratios have the same properties as fractions and therefore, ratio and quotient have almost, but not exactly, the same meaning. For instance, in the indicated operation 6 ÷ 2 =

3, the quotient 3 denotes the number of times 2 is contained in 6. However, since a ratio is the numerical measure of the relation between two similar quantities, it should be stated so as to include both terms. Thus, the ratio of 6 to 2 should be stated as 3 to 1. Nevertheless, common usage has abbreviated the expression of ratios so that only the resulting quotient is stated. Hence, the ratio of 6 to 2 is 3, the ratio of 3 to 5 is 0.6, and so on.

Note that the quantities of a ratio must be expressed in terms of the same unit. The ratio of 1 gallon to 2 quarts is *not* $\frac{1}{2}$. To obtain the ratio of 1 gallon to 2 quarts, either the gallon must be expressed in terms of quarts or the quarts in terms of a gallon. Since 1 gallon is equal to 4 quarts, the ratio of 1 gallon to 2 quarts is $\frac{4}{2} = 2$. If two quantities cannot be expressed in terms of the same unit, no ratio is possible between them.

Example. Find the ratio of: (a) 12 to 6; (b) 7 quarts to 21 gallons; (c) 20 cents to 3 dollars.

Solution.

(a) $\frac{12}{6} = 2$. *Ans.*

(b) Since 1 gallon = 4 quarts, then 21 gallons = 84 quarts, so that the ratio is $\frac{7}{84} = \frac{1}{12}$. *Ans.*

(c) Since 3 dollars is equal to 300 cents, the ratio is $\frac{20}{300} = \frac{1}{15}$. *Ans.*

EXERCISE 37

Find the ratio of each of the following.

1. 20 to 4   **2.** 7 to 8   **3.** 27 to 36

4. 35 to 15   **5.** $5\frac{5}{8}$ to $7\frac{1}{2}$ **6.** $2\frac{11}{12}$ to $1\frac{2}{3}$

7. 64 pounds to 1 ton (1 ton = 2000 pounds)

8. 32 acres to 1 square mile (640 acres = 1 square mile)

9. In wrought iron castings, the proportion of aluminum recommended is $3\frac{1}{2}$ ounces to 100 pounds of iron. What is the ratio of aluminum to iron? (16 ounces = 1 pound)

10. A restaurant with 180 available seats has an average patronage of 828 customers daily. What is the ratio of customers to seats available?

8.3. Proportion. A proportion is a statement of the equality of two ratios. Therefore a proportion, by definition, is an equation. Thus, $2:6 = 1:3$ or $\frac{2}{6} = \frac{1}{3}$ is a proportion. In general, $a:b = c:d$ or $\frac{a}{b} = \frac{c}{d}$ is read "a is to b as c is to d."

The outer terms, that is, the first and fourth terms, are called the *extremes*. The extremes in the above proportions are 2 and 3 or a and d. The inner terms or the second and third terms are called the *means*. Hence, 6 and 1 or b and c are the means.

8.4. Fundamental Principle. Consider the proportion $\frac{a}{b} = \frac{c}{d}$ and let r equal the ratio. That is, let $\frac{a}{b} = r$ and since $\frac{a}{b} = \frac{c}{d}$, then $\frac{c}{d} = r$ also. Now if $a \div b = r$ then, by the definition of division, Section 4.1, $a = b \times r$ and similarly, if $c \div d = r$, then $c = d \times r$. Hence, the proportion $a:b = c:d$ becomes $b \times r:b = d \times r:d$, so that the extremes are $b \times r$ and d and the means are b and $d \times r$. Since the product of the means is $b \times (d \times r) = b \times d \times r$ and the product of the extremes is $(b \times r) \times d = b \times d \times r$, the two products are equal. That is,

The product of the means is equal to the product of the extremes.

In symbols, if $a:b = c:d$ or $\frac{a}{b} = \frac{c}{d}$, then $ad = bc$.

This fundamental principle of proportion is called the *rule of three*. Note that if the ratios are written in the form of fractions, the cross products are equal. Thus, if $\frac{a}{b} = \frac{c}{d}$, then $a \times d = b \times c$. Moreover, if $a \times d = b \times c$ and we let $a \times d = x$, then since $a \times d = b \times c$ it follows that $b \times c = x$ also. But if $a \times d = x$, then, by the definition of division, $a = \frac{x}{d}$ and since $b \times c = x$ then, $a = \frac{b \times c}{d}$.

Similarly, it may be shown that $b = \frac{a \times d}{c}$; $c = \frac{a \times d}{b}$; and $d = \frac{b \times c}{d}$. Therefore, if any three terms of a proportion are given, the fourth may be found.

Example. Find the missing number in each of the following proportions:

(a) $\frac{8}{12} = \frac{20}{d}$; (b) $\frac{5}{6} = \frac{c}{12}$; (c) $\frac{a}{20} = \frac{3}{4}$; (d) $\frac{10}{b} = \frac{40}{68}$.

Solution.

(a) If $\frac{8}{12} = \frac{20}{d}$, then $8 \times d = 20 \times 12$ or $8 \times d = 240$ and

$d = \frac{240}{8} = 30$. *Ans.*

(b) If $\frac{5}{6} = \frac{c}{12}$, then $5 \times 12 = 6 \times c$ or $60 = 6 \times c$ and

$\frac{60}{6} = c = 10$. *Ans.*

(c) If $\frac{a}{20} = \frac{3}{4}$, then $a \times 4 = 3 \times 20$ or $a \times 4 = 60$ and

$a = \frac{60}{4} = 15$. *Ans.*

(d) If $\frac{10}{b} = \frac{40}{68}$, then $10 \times 68 = 40 \times b$ or $680 = 40 \times b$ and

$\frac{680}{40} = b = 17$. *Ans.*

8.5. Simple Proportion. Simple proportion, or the comparison of two simple ratios, can be used to solve many practical problems. For example, suppose that an automobile consumes 3 gallons of gasoline in 54 miles; how many gallons of gasoline are needed for a trip of 1,026 miles?

Other things being equal, the number of gallons required for a trip of 1,026 miles *bears the same relation* to the number of gallons, 3, required for a trip of 54 miles. Hence, if a denotes the number of gallons of gasoline required for a trip of 1,026 miles, we have the proportion $a:3 = 1,026:54$ or $\frac{a}{3} = \frac{1,026}{54}$. Then $a \times$

$54 = 1,026 \times 3$ or $a \times 54 = 3,078$, and $a = \frac{3,078}{54} = 57$ gallons.

It is important to note that the quantities must be in proportion, that is, the ratios of the quantities to each other must be equal and *remain constant*.

For example, if a stone dropped from the Empire State Building, 1,250 feet high, will hit the ground in 8.8 seconds, how high is the Waldorf-Astoria Hotel, if the same stone hits the ground 6.3 seconds after being dropped? The answer is *not* given by the proportion $\dfrac{8.8}{6.3} = \dfrac{1,250}{d}$, for in this case, $d = \dfrac{1,250 \times 6.3}{8.8} = 894.89$

feet, whereas the correct answer is 625 feet. A proportion cannot be used here, because the distance is not the same for the first second of the fall as for every succeeding second; thus, the quantities are not in proportion.

Example.   If silver coin is an alloy of silver and copper in the ratio of 37 to 3, how many pounds of copper are necessary to alloy 481 pounds of silver?

Solution.   Let a represent pounds of copper. Then, $\dfrac{a}{481} = \dfrac{3}{37}$ so that

$a \times 37 = 3 \times 481$ or $a \times 37 = 1,443$; hence,

$$a = \dfrac{1,443}{37} = 39 \text{ lbs.} \quad Ans.$$

EXERCISE 38

Find the unknown term in each of the following proportions.

1. $16:8 = c:5$
2. $25:15 = 10:d$
3. $13:b = 39:96$
4. $a:25 = 6:50$
5. $1:2 = 3:d$
6. $3:8 = c:19$
7. $a:32 = 105:224$
8. $15:b = 5:1$
9. $4:2 = c:8$

10. A water meter registers 450 cubic feet in 30 days. If the ratio remains constant, how many cubic feet will it register in 210 days?

11. If a house assessed for $9,000 must pay $198 in taxes, how much should a house assessed for $10,500 pay?

12. A pole 6 feet high casts a shadow 8 feet long. How high is a tree which, at the same time, casts a shadow 75 feet long?

13. In ordinary foundation work, engineers specify a 1:3:6 mixture of concrete, that is 1 part cement, 3 parts sand, and 6 parts gravel. How many pounds of each are required to make a 2,500 pound mixture?

14. Four merchants jointly shipped some goods whose freight charges were $75,800. How much did each pay, if the first shipped 240 barrels, the second 190, the third 90, and the fourth 80?

15. On a certain map, a line $2\frac{1}{16}$ inches long represents 15 miles. If the distance from Albany, New York to Boston, Massachusetts is represented by a line $25\frac{3}{10}$ inches long, what is the distance from Albany to Boston?

8.6. Properties of Proportion.   The proof of the fundamental principle of proportion (Section 8.4) shows that in order for two ratios to be in proportion, the product of the means *must* equal the product of the extremes. Hence, the proportion may be expressed in any form whatever as long as these products are equal. If $a:b = c:d$, there are eight different ways in which the proportion may be expressed without changing the fact that $a \times d = b \times c$.

1) Given . . . . $a:b = c:d$
2) Interchanging the means in 1 . . . . $a:c = b:d$
3)   "   the extremes in 1 . . . . $d:b = c:a$
4)   "   the means in 3 . . . . $d:c = b:a$
5)   "   the ratios in 1 . . . . $c:d = a:b$
6)   "   the ratios in 2 . . . . $b:d = a:c$
7)   "   the ratios in 3 . . . . $c:a = d:b$
8)   "   the ratios in 4 . . . . $b:a = d:c$

All of these are different forms of the same proportion, for in each case $a \times d = b \times c$. The second form shows that if four numbers are in proportion, they are in proportion by *alternation;* that is, the first term is to the third term as the second is to the fourth. The eighth form shows that the numbers are also in proportion by *inversion;* that is, the second term is to the first term as the fourth is to the third. For example, if $7:5 = 14:10$, then by (2), $7:14 = 5:10$, and by (8), $5:7 = 10:14$.

Now consider the proportion $a:b = a:c$ and let r equal the ratio. Since $\frac{a}{b} = r = \frac{a}{c}$, it follows by the definition of division that $a = b \times r$ and $a = c \times r$. Hence, $b \times r = c \times r$, for they are both equal to a, and evidently $b = c$. Therefore,

9) *If the first and third terms of a proportion are equal, the second and fourth terms are equal.*

Similarly it may easily be proved that,

10) *If the first and second terms of a proportion are equal, the third and fourth terms are equal.*

For example, if $5:8 = 5:d$, then by (9), $d = 8$; and if $8:8 = c:d$, then by (10), $c = d$.

Now consider the two proportions $a:b = c:d$ and $a:b = c:e$, or $\frac{a}{b} = \frac{c}{d}$ and $\frac{a}{b} = \frac{c}{e}$. It immediately follows that $\frac{c}{d} = \frac{c}{e}$, for they are each equal to $\frac{a}{b}$. But if $\frac{c}{d} = \frac{c}{e}$ then by (9), $d = e$. Therefore,

11) *If three terms of a proportion are equal to the corresponding three terms of another proportion, the fourth terms of the two proportions are equal.*

Thus, if $3:5 = 9:15$ and $3:5 = 9:d$, then by (11), $d = 15$.

If in a proportion $a:b = c:d$, $b = c$, it may be written $a:b = b:d$ and the number b is called the *mean proportional* between a and d. But if $\frac{a}{b} = \frac{b}{d}$, then by the fundamental principle (Section 8.4), $a \times d = b^2$ and $b = \sqrt{a \times d}$. Hence,

12) *The mean proportional between two numbers is the positive square root of their product.*

For example, the mean proportional between 2 and 8 is equal to $\sqrt{8 \times 2} = \sqrt{16} = 4$.

If in the proportion $a:b = c:d$ or $\frac{a}{b} = \frac{c}{d}$, 1 is added to each side, the equality remains unchanged. Hence, if $\frac{a}{b} = \frac{c}{d}$, then $\frac{a}{b} + 1 = \frac{c}{d} + 1$ and $\frac{a+b}{b} = \frac{c+d}{d}$, that is, $(a + b):b = (c + d):d$.

Similarly, by using form (8) and adding 1 to each side, it can be shown that if $a:b = c:d$, then $(a + b):a = (c + d):c$. Therefore,

13) *If four numbers are in proportion, they are in proportion by addition; that is, the sum of the first two terms is to the second (or first) term as the sum of the last two terms is to the fourth (or third) term.*

Thus, if $2:5 = 6:15$, then $(2 + 5):5 = (6 + 15):15$; that is, $7:5 = 21:15$. Moreover, if $2:5 = 6:15$, then $(2 + 5):2 = (6 + 15):6$; that is, $7:2 = 21:6$.

EXERCISE 39

(a) Express each of the following proportions in eight different ways.
(b) Show (in two ways) that each of the following proportions is in proportion by addition.

1. $8:2 = 16:4$
2. $2:1 = 10:5$
3. $3:6 = 1:2$
4. $25:15 = 10:6$
5. $3:25 = 6:50$
6. $16:8 = 10:5$
7. $75:160 = 15:32$
8. $51:57 = 85:95$
9. $39:51 = 65:85$

Find the mean proportional between each of the following.

10. 2 and 32
11. 81 and 4
12. 8 and 32
13. 3 and 75
14. 49 and 4
15. 63 and 7
16. 8 and 50
17. 44 and 11
18. 4 and 289

19. A pipe pours 201 gallons of water into a tank in $1\frac{1}{2}$ hours. How long will it take to fill an empty tank with a capacity of 4,623 gallons?

20. An oak beam 9 feet long weighs 69 pounds. What is the weight of an oak beam of the same width that is 6 feet long?

8.7. Inverse Proportion. So far, we have limited our discussion of proportion to direct relationships where one quantity increased as the other quantity increased or one quantity decreased as the other also decreased. Thus, 3 gallons of gasoline were needed to make a trip of 54 miles and 57 gallons were needed for a trip of 1,026 miles. As the number of miles increased the number of gallons needed increased. In an inverse relationship, as one quantity increases, the other decreases or as one quantity decreases, the other increases. Thus, an *inverse proportion* is an equality of two ratios such that two quantities of the same kind are to each other inversely as the two other quantities. For example, consider two driving wheels connected by a belt and let the diameter of the larger wheel be 7 inches. Then the smaller the diameter of the other wheel, the larger the number of revolutions per minute it will make. Thus, the number of revolutions per minute of the wheels is inversely proportional to their respective diameters. Hence, if the larger wheel makes 136 revolutions per minute and the diameter of the smaller wheel is 4 inches and if we let a denote the number of revolutions per minute of this smaller wheel, the number of revolutions per minute of the smaller wheel is to 136, *not* as 4 inches is to 7 inches, but as 7 inches is to 4 inches; that is, inversely as the order indicated by the terms of the first ratio. Hence, $a{:}136 = 7{:}4$ or $4a = 136 \times 7$; that is, $4a = 952$ and $a = \frac{952}{4} = 238$ revolutions.

Example. A train traveling at 40 miles per hour will take 8 hours to cover the distance between two cities. How long will it take a plane traveling at 300 miles per hour to cover the same distance?

Solution. Let a = the time, in hours, that it will take the plane. Since the *greater* the speed, the *smaller* the amount of time, the equalities are inversely in proportion. Thus, $a{:}8 = 40{:}300$; $300a = 8 \times 40$ or $300a = 320$ and $a = \frac{320}{300} = 1\frac{1}{15}$ hours or 1 hour and 4 minutes. *Ans.*

8.8. Compound Proportion. The product of two or more simple ratios is called a *compound ratio*. It may be expressed as the product of two or more fractions, such as $\frac{2}{5} \times \frac{3}{4}$, or by the expression $\begin{Bmatrix} 2{:}5 \\ 3{:}4 \end{Bmatrix}$. The statement of the equality of a compound ratio to a simple ratio or of two compound ratios, is called a *compound proportion*. Thus, $3:10 = \begin{Bmatrix} 2:5 \\ 3:4 \end{Bmatrix}$ $\left(\text{that is, } \dfrac{3}{10} = \dfrac{2}{5} \times \dfrac{3}{4} \right)$ and $\begin{Bmatrix} 4:5 \\ 3:8 \end{Bmatrix} = \begin{Bmatrix} 2:5 \\ 3:4 \end{Bmatrix}$ $\left(\text{that is, } \dfrac{4}{5} \times \dfrac{3}{8} = \dfrac{2}{5} \times \dfrac{3}{4} \right)$ are compound proportions.

Now consider the compound proportion $\begin{Bmatrix} a:b \\ c:d \end{Bmatrix} = \begin{Bmatrix} m:n \\ v:w \end{Bmatrix}$ or $\dfrac{a}{b} \times \dfrac{c}{d} = \dfrac{m}{n} \times \dfrac{v}{w}$. Performing the indicated multiplication we obtain $\dfrac{a \times c}{b \times d} = \dfrac{m \times v}{n \times w}$. Each one of these products can be considered as a term and hence, by the fundamental principle (Section 8.4), $a \times c \times n \times w = b \times d \times m \times v$. That is,

1) *In a compound proportion the product of all the terms in the means equals the product of all the terms in the extremes.*

Thus, if $\begin{Bmatrix} 4:5 \\ 3:8 \end{Bmatrix} = \begin{Bmatrix} 2:5 \\ 3:4 \end{Bmatrix}$ or $\dfrac{4}{5} \times \dfrac{3}{8} = \dfrac{2}{5} \times \dfrac{3}{4}$, then $4 \times 3 \times 5 \times 4 = 5 \times 8 \times 2 \times 3$.

From this principle it can be deduced that

2) *Any term in the extremes equals the product of all the means divided by the product of the other terms in the extremes.*

Thus, if $\begin{Bmatrix} a:5 \\ 3:8 \end{Bmatrix} = \begin{Bmatrix} 2:5 \\ 3:4 \end{Bmatrix}$, then $a = \dfrac{5 \times 8 \times 2 \times 3}{3 \times 5 \times 4} = 4$.

3) *Any term in the means equals the product of all the extremes divided by the product of the other terms in the means.*

Hence, if $\begin{Bmatrix} 4:5 \\ 3:8 \end{Bmatrix} = \begin{Bmatrix} m:5 \\ 3:4 \end{Bmatrix}$, then $m = \dfrac{4 \times 3 \times 5 \times 4}{5 \times 8 \times 3} = 2$.

To solve a problem by compound proportion use the same principles used in simple proportions.

Example. If 10 men can dig a foundation 12 yards long in 8 days, how many days will it take 5 men to dig a foundation 15 yards long?

Solution. The greater the number of men, the smaller the number of days required. The longer the foundation, the greater the number of days required. Hence, the number of days is in proportion inversely to the number of men, and directly, to the length of the foundation. Thus if we let a denote the number of days required, then $a:8 = \begin{Bmatrix} 10:5 \\ 15:12 \end{Bmatrix}$ and

$a = \dfrac{8 \times 10 \times 15}{5 \times 12} = 20$ days. *Ans.*

EXERCISE 40

1. The driving wheel of an electric motor is 4 inches in diameter and makes 2,250 revolutions per minute. If this driving wheel is connected by a belt to another wheel 5 inches in diameter, how many revolutions per minute will the second wheel make?

2. If it takes 2 workmen 20 days to do a certain job, how long will it take 5 workmen to do the same job?

3. A rectangular field must have an area of one acre. The length of the field varies inversely as its width. If the length is 121 yards when the width is 40 yards, what will the length be when the width is 11 yards?

4. It takes a plane whose speed is 250 miles per hour 5 hours and 36 minutes to cover a certain distance. What is the speed of a plane that covers the same distance in 3 hours and 30 minutes?

5. The velocities of the flow of water through pipes running full, and receiving water from an unlimited source, varies inversely as the areas of the cross sections. If for a pipe whose cross-section area is $28\frac{2}{7}$ square inches, the velocity is 2 feet per second, what is the velocity for a pipe whose cross-section area is $3\frac{1}{7}$ square inches?

6. A hotel manager estimates the food consumed by 1,500 guests, eating 3 meals a day for 8 days, costs $16,800. How much will it cost to serve 3,200 guests, eating 2 meals a day, for 10 days?

7. If $3,850 yields $231 in 2 years, how much will $23,100 yield in 11 years?

8. If 20 men earn $6,080 in 19 days, how much should 33 men earn in 7 days?

9. Working on the basis of 8 hours a day and a 6-day week, the capacity of a bed factory is 350 beds per week. How long will it take to fill an order for 525 beds if the men work an extra hour each day?

10. A construction company agrees to lay 55 miles of railway in 675 days. If 125 men finish 5 miles in 81 days, how many more men must they engage to finish the work in the given time?

8.9. Per Cent. The term *per cent*, from the Latin "per centum," means by the hundred. Thus, *rate per cent* is defined as the ratio of two similar quantities expressed in hundredths. For example, 6 per cent means 6 out of every hundred or $\frac{6}{100}$. The symbol %, read per cent, may be used, instead of the decimal fraction, to represent the ratio of two similar quantities expressed in hundredths. For example, 6% is read 6 per cent; it means 6 out of every hundred and is equivalent to $\frac{6}{100}$. That is, the symbol % can be thought of as replacing the denominator 100. Thus, $0.05 = \frac{5}{100} = 5\%$, $0.25 = \frac{25}{100} = 25\%$, and so on.

8.10. Per Cent Expressed as a Decimal or a Common Fraction.
By definition, 75% is read 75 per cent; it means 75 hundredths
and hence it can be written 0.75. Similarly, 125% is read 125 per
cent and means 125 hundredths. Since 100 hundredths is 1 unit,
125 hundredths is 1 unit and 25 hundredths or 1.25. Hence,
125% = 1.25. Finally, $5\frac{5}{8}\%$ (read $5\frac{5}{8}$ per cent) means $5\frac{5}{8}$ hundredths and hence $5\frac{5}{8}\% = 0.05\frac{5}{8} = 0.05625$. Therefore,

*To express a per cent as a decimal, omit the per cent sign and
move the decimal point two places to the left.*

A rate per cent is, by definition, equivalent to a decimal frac-
tion with a denominator of 100. Thus, $20\% = \frac{20}{100}$. But $\frac{20}{100} = \frac{1}{5}$.
Therefore, $20\% = \frac{1}{5}$. Hence,

*To express a per cent as a common fraction, omit the per cent
sign, write the per cent as the numerator of a fraction whose denominator is 100, and reduce this fraction to its lowest terms.*

Example. Express each of the following per cents as (1) a decimal;
(2) a common fraction: (a) 25%; (b) 135%; (c) $4\frac{1}{2}\%$; (d) $\frac{1}{4}\%$.
Solution.

(a) (1) $25\% = 0.25$; (2) $25\% = \frac{25}{100} = \frac{1}{4}$;

(b) (1) $135\% = 1.35$; (2) $135\% = \frac{135}{100} = \frac{27}{20} = 1\frac{7}{20}$;

(c) (1) $4\frac{1}{2}\% = 0.04\frac{1}{2} = 0.045$; (2) $4\frac{1}{2}\% = \frac{4\frac{1}{2}}{100} = \frac{9}{200}$;

(d) (1) $\frac{1}{4}\% = 0.00\frac{1}{4} = 0.0025$; (2) $\frac{1}{4}\% = \frac{\frac{1}{4}}{100} = \frac{1}{400}$.

**8.11. Decimal Fractions and Common Fractions Expressed as a
Per Cent.** In the decimal notation, the digit written in the second
place to the right of the units place indicates hundredths (Section
7.2). Since per cent means hundredths, it follows that the first
two digits to the right of the units place give per cent. For example, 0.35 or 35 hundredths is 35%. Similarly, 1.25 is 125 hundredths or 125%. Moreover, 0.045 means 4 hundredths and 5
thousandths. But 5 thousandths is 5 tenths (0.5) of one hundredth. Hence 0.045 is 4 and 5 tenths hundredths or 4.5 hundredths or 4.5%. Therefore,

*To express a decimal as a per cent, move the decimal point two
places to the right, and write the per cent sign after the number.*

As shown in Section 7.9, any common fraction can be written

as a decimal. Any decimal, as shown on the preceding page, can be expressed as a per cent. Therefore, any common fraction can be written as a per cent. For example, $\frac{7}{20}$ = 0.35 and 0.35 = 35%; so, $\frac{7}{20}$ = 35%. Similarly, $\frac{5}{4}$ = 1.25 and 1.25 = 125%; hence, $\frac{5}{4}$ = 125%. Finally, $\frac{9}{200}$ = $\frac{45}{1,000}$ = 0.045 and 0.045 = 4.5%; hence, $\frac{9}{200}$ = 4.5%. Therefore,

To express a common fraction as a per cent, change the common fraction to a decimal, then express this decimal as a per cent.

Example. Express each of the following as a per cent: (a) 0.35; (b) 1.15; (c) 0.075; (d) 0.0075; (e) $\frac{3}{20}$; (f) $\frac{9}{8}$; (g) $\frac{1}{40}$; (h) $\frac{1}{250}$.

Solution.

(a) 0.35 = 35%;   (b) 1.15 = 115%;

(c) 0.075 = 7.5% = $7\frac{1}{2}$%;   (d) 0.0075 = 0.75% = $\frac{3}{4}$%;

(e) $\frac{3}{20}$ = 0.15 = 15%;   (f) $\frac{9}{8}$ = 1.125 = 112.5% = $112\frac{1}{2}$%;

(g) $\frac{1}{40}$ = 0.025 = 2.5% = $2\frac{1}{2}$%;

(h) $\frac{1}{250}$ = 0.004 = 0.4% = $\frac{2}{5}$%.

EXERCISE 41

Express each of the following per cents as (a) a decimal; (b) a common fraction.

1. $6\frac{1}{4}$%   2. 8%   3. 40%   4. $12\frac{1}{2}$%   5. $37\frac{1}{2}$%

6. $62\frac{1}{2}$%   7. $14\frac{2}{7}$%   8. $\frac{5}{12}$%   9. $\frac{4}{9}$%   10. $33\frac{1}{3}$%

11. $1\frac{3}{4}$%   12. $6\frac{4}{5}$%   13. $3\frac{3}{4}$%   14. $16\frac{3}{8}$%   15. $17\frac{3}{16}$%

Express each of the following as a per cent.

16. 0.45   17. 0.375   18. 0.125   19. 0.0625   20. 0.00875

21. 1.014   22. 1.625   23. $\frac{11}{20}$ 24. $\frac{3}{5}$ 25. $\frac{7}{8}$

26. $\frac{1}{20}$ 27. $\frac{7}{32}$ 28. $\frac{5}{16}$ 29. $\frac{43}{200}$ 30. $\frac{99}{500}$

8.12. Computations Using Per Cents. Computations with per cents involve three distinct quantities: the *base*, the *rate*, and the *percentage*.

The *base* is the quantity on which the percentage is computed. The *rate per cent* is the ratio of the percentage to the base expressed in hundredths, that is, the number of hundredths of the base. The *percentage* is the result obtained by taking a number of hundredths of the base. For example, a house assessed at $50,000 pays 3% of the assessed value in taxes, that is, $1,500. In this case, $50,000 is the base, 3% is the rate per cent, and $1,500 is the percentage.

Note that the rate per cent, 3%, is a ratio and therefore an abstract number, whereas the percentage, $1,500, is an amount expressed in the same unit as the base.

8.13. Finding the Percentage. If the base and the rate per cent are given, a general rule for finding the percentage can be derived as follows. By definition, rate per cent is the number of hundredths of the base. Thus, 20% of 400 means 20 hundredths of 400, that is, 20 for every hundred units in 400. Since there are 4 units of 100 each in 400, 20% of 400 is $20 \times 4 = 80$. Now 20% may be expressed as 0.20 or as $\frac{20}{100} = \frac{1}{5}$. If 20% is written 0.20, then $400 \times 0.20 = 80$. If 20% is written $\frac{1}{5}$, then $400 \times \frac{1}{5} = 80$. Therefore,

To find the percentage:

1) *Reduce the rate per cent to a decimal or to a common fraction.*
2) *Multiply this decimal, or common fraction, by the base.*

Example. W. Johnson owns 300 shares of common stock with a par value of $100 a share. If the directors declare a 5% dividend, find the amount Johnson will receive in dividends.

Solution. The total par value of Johnson's shares is $300 \times \$100 =$ $30,000. Since $5\% = 0.05$, or $5\% = \dfrac{5}{100} = \dfrac{1}{20}$, then

$$\$30,000 \times 0.05 = \$1,500. \quad Ans.$$

$$\text{Or} \quad \$30,000 \times \frac{1}{20} = \$1,500. \quad Ans.$$

8.14. Finding the Rate. By definition, rate per cent is the ratio of two similar quantities expressed as hundredths. The ratio of one quantity to another similar quantity is the result obtained by dividing the first quantity by the second (Section 8.1). Hence, to express the ratio of two similar quantities, such as $2 and $16, as a rate per cent, divide $2 by $16 and then express the result in hundredths. Thus, $\frac{2}{16} = \frac{1}{8} = 0.125$, and $0.125 = 12.5\% = 12\frac{1}{2}\%$; hence, $2 is $12\frac{1}{2}\%$ of $16. Therefore,

> *To find the rate, divide the percentage by the base and express this quotient as a rate per cent.*

Example. J. Walker bought a house for $30,000. If his net annual rent from this house is $1,350, what rate per cent of his investment does Mr. Walker receive annually?

Solution. Here the base is $30,000 and the percentage is $1,350; hence,

$$\text{rate} = \frac{1,350}{30,000} = 0.045 = 4.5\% = 4\frac{1}{2}\%. \quad Ans.$$

8.15. Finding the Base. If the rate per cent and the percentage are given, a rule to find the base can be derived by reasoning as follows. Assume that a number is to be found of which 86 is 40%. Here 40% is the rate, 86 is the percentage, and the base is to be found. Since $40\% = \frac{40}{100} = \frac{2}{5}$, then $\frac{2}{5}$ of the number is 86. But if $\frac{2}{5}$ of the number is 86, then $\frac{1}{5}$ is $\frac{86}{2}$ and the number is $5 \times \frac{86}{2} = 215$.

Similarly, if 40% of a number is 86, then 1% of the number is $\frac{86}{40}$, and 100% of the number is $\frac{86}{40} \times 100 = \frac{86}{0.40} = 215$.

Note that if the rate per cent, expressed as a fraction, is divided into the percentage, the base is obtained. Thus, $86 \div \frac{2}{5} = 86 \times \frac{5}{2} = 215$. Similarly, if the rate per cent expressed as a decimal is divided into the percentage, the base is obtained. Thus, $86 \div 0.40 = 215$. Therefore,

> *To find the base:*
> 1) *Express the rate per cent as a common fraction or as a decimal.*
> 2) *Divide the common fraction or decimal into the percentage.*

Example. A merchant allows a 15% discount from the marked price. If the discount amounts to $13.50, what is the marked price of this article?

Solution. Here, $13.50 is the percentage, 15% is the rate, and the marked price is the base. Now $15\% = 0.15$ or $\frac{15}{100} = \frac{3}{20}$; hence,

$$\text{marked price} = \$13.50 \div 0.15 = \$90. \quad Ans.$$

$$\text{Or marked price} = \$13.50 \div \frac{3}{20} = \$13.50 \times \frac{20}{3} = \$90. \quad Ans.$$

EXERCISE 42

Calculate the per cent.

1. 30% of 72   2. 20% of 54
3. 15% of 125   4. $6\frac{1}{4}$% of $300
5. $12\frac{1}{2}$% of $350   6. $3\frac{3}{4}$% of 112.5
7. $1\frac{1}{2}$% of 500   8. $\frac{5}{12}$% of 780

Calculate the rate per cent to the nearest tenth of a per cent.

9. Percentage 50, base 250   10. Base 310, percentage 93
11. Percentage 12.43, base 310.75   12. Percentage 40.61, base 564
13. What per cent of 648 is 81?   14. 34.91 is what per cent of 149.36?

Calculate the base.

15. Percentage 45, rate 30%   16. Rate $8\frac{1}{2}$%, percentage 187
17. Percentage $1,615, rate 60%   18. Percentage $209, rate $5\frac{1}{2}$%
19. 735 is $\frac{7}{8}$% of what number?   20. $89 is $2\frac{1}{2}$% of what amount?

21. A television set is marked at $640 less $14\frac{2}{7}$%. Find the net price.

22. A man borrowed $750 and agreed to pay 6% per year as interest. Find the interest paid for 4 months.

23. A man earns a salary of $9,500 a year. He spends 20% for rent, 25% for food, 10% for clothing, 5% for doctors and medicines, 2% for insurance, 9% for amusements, 19% for taxes and miscellaneous items, and saves the rest. (a) How much does he spend for each item? (b) How much does he save?

24. A solution containing $7\frac{1}{2}$ pounds of salt, 18 pounds of sodium carbonate, and 6 pounds of silver chloride is used to silver plate an object. Find the rate per cent of (a) salt used; (b) sodium carbonate used; (c) silver chloride used.

25. The daily average number of customers in a bakery shop increased from 35 to 64. What is the rate per cent of increase in the number of customers?

26. The payments on a loan of $100 for one year increased from $8.87 per month to $8.91 per month. Find, to the nearest hundredth of a per cent, the rate per cent of increase.

27. A man invested 40% of his capital in stocks. If the amount invested was $56,384, what was his total capital?

28. A retailer advertised a discount of 20% on television sets. Taking advantage of this discount, a man paid $239.60 for a television set. Find the original marked price.

29. After a cut of 5%, the wages of a certain man were $4.37 per hour. What were the man's wages before the cut?

30. After the expenses of a business increased by $3\frac{2}{5}$%, the expenses amounted to $98,230. What were the expenses before the increase?

8.16. Applications of Percentage.* The great convenience of computing by per cent makes its applications in business extensive. Some of these applications are briefly discussed in the following paragraphs.

Commission, or *brokerage*, is the amount paid to an agent, or broker, who buys or sells goods for someone else. Usually, the commission is expressed as a rate per cent of the volume of business.

Example. An agent sold 1,890 bushels of potatoes at $0.82 a bushel. His charges were: freight—$189.60; storage—$75; commission—4%. Find the net proceeds.

Solution.

1,890 bushels at $0.82 . . . . $1,549.80
Freight $189.60
Storage 75.00
Commission $1,549.80 × 0.04. . . . . 61.99
Total charges 326.59
Net proceeds . . . . $1,223.21. *Ans.*

Example. A broker sold corn at $2.97 a bushel, charging a commission of 5%. His commission amounted to $950.40. How many bushels did he sell?

Solution. Since 5% of sales = commission, then

$$5\% \text{ of sales} = \$950.40$$

$$1\% \text{ of sales} = \frac{\$950.40}{5}$$

$$\text{sales} = \frac{\$950.40}{5} \times 100 = \$19,008.$$

Hence, $19,008 ÷ $2.97 = 6,400 bushels. *Ans.*

A *commercial discount* is any deduction allowed from the marked price or list price of an article. Commercial discounts are either (1) trade discounts or (2) cash discounts.

1) A *trade discount* is a deduction allowed from the marked price or list price of an article for any reason whatever except to induce payment within some specified time. In order to keep the price within competitive market prices and avoid the necessity of publishing new catalogues, several successive discounts can be offered from the catalogue price. *Each successive discount is computed as a percentage of the remainder left after deducting the*

**For an extensive treatment of percentage see Mira and Hartmann, *Business Mathematics* (Princeton: Van Nostrand, 1955).

previous discount. For example, a marked price of $100 less 20%
and 10% gives a net price of $72 computed as follows:

$$
\begin{aligned}
\text{First discount} &\dots\dots\dots\ \$100 \times 0.20 = \$20 \\
\text{First remainder} &\ \dots\dots\dots\ \$100 - \$20 = \$80 \\
\text{Second discount} &\dots\dots\dots\ \$80 \times 0.10 = \$8 \\
\text{Net price} &\dots\dots\dots\dots\ \$80 - \$8 = \$72
\end{aligned}
$$

2) A *cash discount* is a deduction allowed from the net price
as an inducement for immediate payment or for payment within
a specified time. "Cash 5%" means that 5% may be deducted
from the net price provided payment is made at once. The expression "terms $5/10, n/60$" means that 60 days after the date of
the invoice the net price of the merchandise is due and payable,
but that if payment is made within 10 days of the date of the invoice, a discount of 5% of the net price of the merchandise is
allowed. Thus, if the above article listed at $100 less 20% and
10% is bought May 15, the terms are $5/10, n/60$, and the invoice
is paid May 25, the amount to be paid, that is, the net amount of
the invoice, is the net price, $72, less 5%. Thus, cash discount =
$72 × 0.05 = $3.60 and net amount of invoice = $72 − $3.60 =
$68.40.

Example. A television set is listed at $650 less 20% and 10%, terms
$3/10, n/30$. This set is bought April 10 and the invoice is paid April 20.
Find the net amount of the invoice.

Solution.
$$
\begin{aligned}
\text{First discount} &\dots\dots\dots\ \$650 \times 0.20 = \$130. \\
\text{First remainder} &\dots\dots\ \$650 - \$130 = \$520. \\
\text{Second discount} &\dots\dots\ \$520 \times 0.10 = \$52. \\
\text{Net price} &\dots\dots\dots\ \$520 - \$52 = \$468. \\
\text{Cash discount} &\dots\dots\dots\ \$468 \times 0.03 = \$14.04. \\
\text{Net amount of invoice} &\dots\dots\ \$468 - \$14.04 = \$453.96. \quad \textit{Ans.}
\end{aligned}
$$

This solution may be simplified as follows: A discount of 20% from the
list price leaves a first remainder of 80% of the list price. A discount of
10% of the first remainder leaves 90% of the first remainder as a net
price, that is, 90% of 80% of the list price. But $90\% = \frac{90}{100} = \frac{9}{10}$ and
$80\% = \frac{80}{100} = \frac{4}{5}$. Hence 90% of 80% of the list price is $\frac{9}{10}$ of $\frac{4}{5}$ of $650,
that is

$$\$650 \times \tfrac{4}{5} \times \tfrac{9}{10} = 13 \times 4 \times 9 = \$468 \text{ net price.}$$

A cash discount of 3% from the net price leaves 97% of the net price as
the net amount of the invoice. Since $97\% = \frac{97}{100} = 0.97$, then

$$\$468 \times 0.97 = \$453.96. \quad \textit{Ans.}$$

Example. Find the single discount equivalent to the discount series 20%, 10%, and 5%.

Solution. A discount of 20% from the list price leaves

$$\text{First remainder} = 80\% \text{ of the list price.}$$

A discount of 10% from the first remainder leaves

$$\text{Second remainder} = 90\% \text{ of the first remainder or}$$
$$= 90\% \text{ of } 80\% \text{ of the list price.}$$

A discount of 5% from the second remainder leaves

$$\text{Net price} = 95\% \text{ of the second remainder or}$$
$$= 95\% \text{ of } 90\% \text{ of } 80\% \text{ of the list price.}$$

But 95% = 0.95, 90% = 0.90, and 80% = 0.80; hence,

$$\text{Net price} = 0.95 \times 0.90 \times 0.80 \text{ of the list price or}$$
$$= 0.684 \text{ of the list price.}$$

But 0.684 = 68.4%; hence, the net price is 68.4% of the list price. If the net price is 68.4% of the list price, then the discount is

$$100\% - 68.4\% = 31.6\%. \quad Ans.$$

EXERCISE 43

1. A commission merchant sold 1,500 bushels of corn at 97.9 cents per bushel. His charges were: freight, $93.90; storage, $20.15; commission, $3\frac{1}{2}\%$. Find the net proceeds.

2. A broker received instructions to buy $4,000 worth of potatoes and to pay all expenses out of this amount. The agent bought the potatoes at $2.12 per bushel and charged a commission of 4%. (a) How many bushels did he buy? (b) What was the balance?

3. A set of furniture is listed at $950 less $33\frac{1}{3}\%$ and 10%. (a) Find the net price. (b) Find the single discount that would give the same net price.

4. On March 27, Gold & Dewey bought merchandise amounting to $7,100.32 less $6\frac{1}{4}\%$, 10%, and 4%, terms 4/10, n/30. They paid for this merchandise on April 3. (a) Find the net amount of the invoice. (b) Find the single trade discount that will give the same net price.

5. Fraser & Co. buy washing machines for $650 less 25%, 10%, and 5%. What is the selling price if their gross profit is 40% of the net cost?

6. A broker bought 12,500 pounds of hogs at $13.50 per 100 pounds. His charges were: freight, $315.45; inspection, $21.90; commission, 5%. Find the gross cost.

7. A broker bought 288 dozen eggs at 29.6 cents a dozen. His charges were: freight, $45.30; inspection, $3.10; commission, $3\frac{3}{4}\%$. Find the gross cost.

8. An agent sold 3,500 bushels of rye at $92\frac{1}{2}$ cents per bushel. His

charges were: freight, $115; storage, $78; commission, 4%. Find the net proceeds.

9. An agent sold apples at $2.14 a barrel. If he charged a $3\frac{1}{4}\%$ commission, and this commission amounted to $88.19, how many barrels of apples did he sell?

10. A commission merchant bought a herd of cows at $219 per head. If he charged a $2\frac{1}{2}\%$ commission and this commission amounted to $2,321.40, how many cows did he buy?

11. A commission merchant received instructions to buy $1,025 worth of barley and to pay all expenses out of this amount. The agent bought the barley at 84.8 cents per bushel, charged a 3% commission, and paid freight charges of 3 cents a bushel. (a) How many bushels did he buy? (b) What was the balance?

12. Redmond & Co. buy men's hats at $48 a dozen less 20% and 10%, terms 2/10, n/30. (a) Find the net price of a dozen hats. (b) Find the single discount equivalent to the discount series of 20% and 10%.

13. On October 21, Fern & White bought merchandise amounting to $1,491.12 less 5%, 10%, and $12\frac{1}{2}\%$, terms 5/10, 2/30, n/60. Fern & White paid for this merchandise on November 20. (a) Find the net amount of the bill. (b) Find the single trade discount equivalent to 5%, 10%, and $12\frac{1}{2}\%$.

14. On February 26, Fuller & Co. bought merchandise amounting to $1,853.16 less 20%, 13%, and 5%, terms 6/10, 2/30, n/60. They paid for this merchandise on March 27. (a) Find the net amount of the invoice. (b) Find the single trade discount that will give the same net price.

15. On October 25, Smith & Shane bought merchandise amounting to $5,761.50 less 6% and $7\frac{1}{2}\%$, terms 6/10, n/30. They paid for this merchandise on November 4. (a) Find the net amount of the invoice. (b) Find the single trade discount that will give the same net price.

16. Roger & Niles sell rugs listed at $300 less 10% and 5%. Find, to the nearest integral per cent, the additional trade discount that must be allowed to sell these rugs at $239.40 each.

17. Blackston & Co. list a deep freeze for $640 less $12\frac{1}{2}\%$ and 5%. Find, to the nearest integral per cent, the additional trade discount that must be allowed to sell this deep freeze for $478.80.

18. Levin & Clay list an electric drier for $400 less 15% and $6\frac{2}{3}\%$. Find, to the nearest integral per cent, the additional trade discount that must be allowed to sell this drier for $307.81.

19. Palmer & Black buy air conditioners at $333 less $16\frac{2}{3}\%$ and 10%. Find the selling price if their gross profit is (a) 30% of the net cost; (b) 30% of the selling price.

20. Foster, Roth, & Co. buy men's suits at $95 less 10% and 5%. Find the quoted selling price if their gross profit is (a) 40% of the net cost; (b) 40% of the selling price.

Appendix

Check Numbers. *The sum of the check numbers of the addends is equal to the check number of the sum.* Let m and n be any two integers and divide each one by the same integer d. Then

(1) $$m = dq + r_1 \quad \text{and}$$
(2) $$n = dc + r_2$$

where q, c, r_1, and r_2 are integers; q and c represent the respective quotients; and r_1 and r_2 denote each of the remainders, that is, r_1 is the check number of m and r_2 is the check number of n. Adding (1) and (2) we obtain

$$m + n = dq + dc + r_1 + r_2 \quad \text{or}$$
$$m + n = d(q + c) + (r_1 + r_2)$$

Thus, if $m + n$ is divided by d we obtain a quotient $q + c$ and a remainder, $r_1 + r_2$. Therefore, the check number of the sum, $m + n$ is $r_1 + r_2$, which is the sum of the check numbers of the addends.

The product of the check numbers of the factors of a number is equal to the check number of the product. Multiplying (1) by (2) we obtain

$$mn = (dq + r_1) \times (dc + r_2) \quad \text{or}$$
$$mn = d^2cq + dcr_1 + dqr_2 + r_1r_2 \quad \text{and}$$
$$mn = d(dcq + cr_1 + qr_2) + r_1r_2.$$

Thus, if the product mn is divided by d we obtain a quotient, $dcq + cr_1 + qr_2$, and a remainder, r_1r_2. Therefore, the check number of the product, mn, is r_1r_2, which is the product of the check numbers of the factors.

Proof of Divisibility by 9. Any number can be written in the form

$$N = a_nb^n + a_{n-1}b^{n-1} + \cdots + a_2b^2 + a_1b + a_0$$

where the a's represent the digits which make up the number and

b denotes the base of the system used (see page 10). Now if we add and subtract a_n, a_{n-1}, $\ldots$, a_2, a_1, and a_0, the value of the number is unchanged. Thus,

$$N = a_n b^n - a_n + a_{n-1} b^{n-1} - a_{n-1} + \cdots + a_2 b^2 - a_2$$
$$+ a_1 b - a_1 + a_0 + a_1 + a_2 + \cdots + a_{n-1} + a_n.$$

This expression can be written

$$N = a_n(b^n - 1) + a_{n-1}(b^{n-1} - 1) + \cdots + a_2(b^2 - 1)$$
$$+ a_1(b - 1) + a_0 + a_1 + a_2 + \cdots + a_{n-1} + a_n.$$

Now $b^n - 1$, $b^{n-1} - 1$, $\cdots$, $b^2 - 1$, $b - 1$ are each divisible by $b - 1$ for each contains $b - 1$ as a factor. It immediately follows that if a number is divided by $b - 1$, the remainder, if any, will be due to the division of the sum $a_0 + a_1 + a_2 + \cdots + a_{n-1} + a_n$. That is, *any number divided by $b - 1$ leaves the same remainder as the sum of its digits divided by $b - 1$*. Since in our number system we use the base 10, if we let $b = 10$, then $b - 1 = 10 - 1 = 9$. Therefore, any number divided by 9 leaves the same remainder as the sum of its digits divided by 9. It immediately follows, by law 5, Section 5.4, that if the sum of the digits of a number is divisible by 9, then the number is divisible by 9.

Proof of Divisibility by 11. Any number can be written in the form

$$N = a_n b^n + a_{n-1} b^{n-1} + \cdots + a_2 b^2 + a_1 b + a_0$$

or reversing the sum,

$$N = a_0 + a_1 b + a_2 b^2 + a_3 b^3 + a_4 b^4 + a_5 b^5 + \cdots$$

where the a's denote the digits which make up the number and b represents the base of the system used (see page 10). Now if we add and subtract a_1, a_2, a_3, a_4, a_5, $\cdots$, the value of the number is unchanged. Thus,

$$N = a_1 b + a_1 + a_2 b^2 - a_2 + a_3 b^3 + a_3 + a_4 b^4 - a_4 + a_5 b^5$$
$$+ a_5 + \cdots + (a_0 + a_2 + a_4 + \cdots) - (a_1 + a_3 + a_5 + \cdots).$$

This expression can be written

$$N = a_1(b + 1) + a_2(b^2 - 1) + a_3(b^3 + 1) + a_4(b^4 - 1)$$
$$+ a_5(b^5 + 1) + \cdots + (a_0 + a_2 + a_4 + \cdots)$$
$$- (a_1 + a_3 + a_5 + \cdots).$$

Now $b + 1$, $b^2 - 1$, $b^3 + 1$, $b^4 - 1$, $b^5 + 1, \cdots$ are each divisible by $b + 1$, for each contains $b + 1$ as a factor. It immediately follows that if a number is divided by $b + 1$, the remainder, if any, will be due to the division of the difference between $(a_0 + a_2 + a_4 + \cdots) - (a_1 + a_3 + a_5 + \cdots)$ by $b + 1$. That is, any number divided by $b + 1$ leaves the same remainder as the division of the difference between the sum of its digits in the odd places and the sum of its digits in the even places, by $b + 1$. Since in our number system we use a base of 10, if we let $b = 10$, then $b + 1 = 10 + 1 = 11$. Therefore, any number divided by 11 leaves the same remainder as the division of the difference between the sum of its digits in the odd places and the sum of its digits in the even places, by 11. It immediately follows, by law 5, Section 5.4, that if the difference between the sum of the digits in the odd places and the sum of the digits in the even places of any number is divisible by 11, then the number is divisible by 11.

Euclidean Algorithm.* From the definition of division, if any integer n is divided by another integer d not equal to 0,

(1) $$n = dq + r$$

where q and r are also integers, r is less than d and d is less than n.

Assume that a number b divides both n and d, then

$$n = ab \qquad \text{and} \qquad d = mb$$

where a and m are integers. From (1),

$$r = n - dq$$

whence

$$r = ab - mbq = b(a - mq)$$

so that b is a divisor of r. Moreover, if b divides d and r, then

$$d = cb \qquad \text{and} \qquad r = eb$$

where c and e are integers. But

$$n = dq + r$$

whence

$$n = cbq + eb = b(cq + e)$$

so that b is a divisor of n. Hence, any number that divides n and d will also divide d and r. That is, the set of all common divisors of n and d is the same as the set of all common divisors of d and r.

---

*An algorithm is a method of solving a certain type of problem.

Since the greatest common divisor (G. C. D.) is a common divisor, then the G. C. D. of the dividend and the divisor is the same as the G. C. D. of the divisor and the remainder, that is,

(2) $G. C. D. (n \& d) = G. C. D. (d \& r)$

Since in any division the remainder r is less than the divisor d, if we divide the divisor d by the first remainder r, we obtain another remainder r_1 less than r and such that by (2)

$$G. C. D. (d \& r) = G. C. D. (r \& r_1).$$

Repeating this process, we will obtain a series of remainders each less than the preceding one by at least 1. Hence, after most d divisions, we will obtain a remainder of 0. Thus, by successive applications of (2), if r_n denotes the last remainder different from zero, we have

$$G. C. D. (n \& d) = G. C. D. (r \& r_1)$$
$$= \cdots = G. C. D. (r_n \& 0) = r_n.$$

Therefore, the greatest common divisor of n and d is the last remainder different from zero.

Proof of the Accuracy of a Product.   In this proof, the consideration of a special case is more effective than a general discussion that would involve complicated notation.

Let N represent a number of 4 significant figures and M denote a number of 5 significant figures. If N is written in scientific notation and n denotes its significant figures, then $N = n10^3 \pm \frac{1}{2}$, where n lies between 1 and 10. Similarly, if m represents the significant figures of M, then $M = m10^4 \pm \frac{1}{2}$, where m lies between 1 and 10.

The product $M \times N$ cannot be greater than $(n10^3 + \frac{1}{2}) \times (m10^4 + \frac{1}{2}) = mn10^7 + \frac{1}{2}(n + 10m)(10^3) + \frac{1}{4}$, or less than $(n10^3 - \frac{1}{2}) \times (m10^4 - \frac{1}{2}) = mn10^7 - \frac{1}{2}(n + 10m)(10^3) + \frac{1}{4}$.

The product $M \times N$ is not equal to $n10^3 \times m10^4 = mn10^7$, but, disregarding the $\frac{1}{4}$, it may vary by as much as $\frac{1}{2}(n + 10m)(10^3)$.

Since n and m are between 1 and 10, $\frac{1}{2}(n + 10m)$ is less than 100 but greater than 10; that is, it is a number of 2 figures. It then follows that $\frac{1}{2}(n + 10m)(10^3)$ is a number of 5 figures. If $n \times m$ is less than 10, then $n10^3 \times m10^4 = mn10^7$ is a number of 8 figures. If $n \times m$ is greater than 10, then $mn10^7$ is a number of 9 figures.

In either case, $mn10^7 \pm \frac{1}{2}(n + 10m)(10^3)$ shows that the last 5 figures of $mn10^7$ may be unreliable and that the product is correct to at most 4 significant figures. That is, the product has the same number of significant figures as the least accurate of the two factors.

Proof of the Accuracy of a Quotient. Let N represent a number of 4 significant figures and M denote a number of 5 significant figures. Assume that $M \times N = P$. Then from the preceding proof, P has 4 significant figures. But if $M \times N = P$, then $N = P \div M$. Hence, if a number, P, of 4 significant figures is divided by a number, M, of 5 significant figures, the quotient, N, has the same number of significant figures, 4, as the least accurate of the two numbers.

Answers
to the Exercises

EXERCISE

1. $6(10^2) + 4(10) + 9$. **2.** $8(10^2) + 4(10) + 2$. **3.** $3(10^3) + 0(10^2) + 2(10) + 1$. **4.** $7(10^3) + 5(10^2) + 0(10) + 8$. **5.** $1(10^4) + 2(10^3) + 7(10^2) + 6(10) + 4$. **6.** $3(10^5) + 0(10^4) + 1(10^3) + 9(10^2) + 0(10) + 5$. **7.** $6(10^5) + 2(10^4) + 8(10^3) + 0(10^2) + 3(10) + 2$. **8.** $4(10^6) + 3(10^5) + 9(10^4) + 8(10^3) + 2(10^2) + 9(10) + 0$. **9.** 94. **10.** 863. **11.** 51. **12.** 2,415. **13.** 172.
14. 1,235. **15.** 243. **16.** 52. **17.** 1,163. **18.** 1,581. **19.** 516.
20. 83. **21.** 103_7. **22.** 10_7. **23.** 10_{12}. **24.** 20_{12}.
25. L + O + V + E, the E is one of the five letters in eight.
26. (a) 1,523; (b) $00\Sigma75$. **27.** (a) 01000; (b) $00\Sigma10$; (c) 10000.
28. Turn VI upside down obtaining $\wedge$ I. On top of this, put VI, making

$$\frac{\text{VI}}{\wedge\text{I}}$$

29. (a) $\text{VI} + \text{IV} = \text{X}$; (b) $\text{MV} - \overline{\text{T}} = \text{V}$.
30. $\text{L} \diamond \text{V} \equiv$.

EXERCISE 1

1. 5. **2.** 5. **3.** 7. **4.** 8. **5.** 7. **6.** 9. **7.** 9. **8.** 11. **9.** 12.
10. 16. **11.** 11. **12.** 17. **13.** 10. **14.** 11. **15.** 12. **16.** 14.
17. 16. **18.** 16. **19.** 13. **20.** 12. **21.** 13. **22.** 13. **23.** 15.
24. 13. **25.** Σ_{12}. **26.** 10_{12}. **27.** 14_{12}. **28.** 13_{12}. **29.** 18_{12}.
30. 17_{12}. **31.** 17_{12}. **32.** $1\Gamma_{12}$.

EXERCISE 2

1. 79. **2.** 99. **3.** 87. **4.** 99. **5.** 87. **6.** 83. **7.** 98. **8.** 86.
9. 116. **10.** 155. **11.** 145. **12.** 144. **13.** 30. **14.** 20. **15.** 20.
16. 15. **17.** 23. **18.** 25. **19.** 320. **20.** 244. **21.** 264.

22. 2,784.   **23.** 2,988.   **24.** 2,100.   **25.** 104_{12}.   **26.** 183_{12}.
27. $1\Gamma9_{12}$.   **28.** 113_7.   **29.** 110_7.   **30.** 165_7.

31.
$$46 + 29 + 13 + 96 = 184$$
$$43 + 71 + 65 + 84 = 263$$
$$28 + 49 + 17 + 64 = 158$$
$$\underline{98 + 67 + 15 + 65} = \underline{245}$$
$$215 + 216 + 110 + 309 = 850$$

32.
$$43 + 73 + 79 + 26 = 221$$
$$23 + 70 + 42 + 24 = 159$$
$$61 + 30 + 69 + 38 = 198$$
$$\underline{18 + 32 + 96 + 35} = \underline{181}$$
$$145 + 205 + 286 + 123 = 759$$

33.
$$62 + 87 + 98 + 24 = 271$$
$$12 + 56 + 97 + 43 = 208$$
$$79 + 28 + 23 + 99 = 229$$
$$\underline{26 + 74 + 13 + 38} = \underline{151}$$
$$179 + 245 + 231 + 204 = 859$$

34.
$$44 + 24 + 79 + 96 = 243$$
$$41 + 76 + 76 + 37 = 230$$
$$94 + 52 + 81 + 99 = 326$$
$$\underline{63 + 48 + 77 + 65} = \underline{253}$$
$$242 + 200 + 313 + 297 = 1,052$$

EXERCISE 3

1. 22.   **2.** 30.   **3.** 30.   **4.** 150.   **5.** 138.   **6.** 136.   **7.** 170.
8. 15,895.   **9.** 20,214.   **10.** 14,270.   **11.** 12,220.   **12.** 11,020.
13. 12 cents. Cut each link of one of the pieces and connect four of the other pieces, then cut two more links and connect the remaining pieces.
14. Take the cat over, return and take the mouse over, bring back the cat, take the dog over, and then return for the cat.
15. 7 days.
16. Neither. But 9 and 5 are 14, and 9 and 5 is 14 are both correct.
17. If 366 have different birthdays, then the 367th must duplicate one of these. If the 366 do not have different birthdays, then at least two must have the same one.

EXERCISE 4

1. 3.   **2.** 1.   **3.** 0.   **4.** 2.   **5.** 5.   **6.** 2.   **7.** 3.   **8.** 3.   **9.** 5.
10. 8.   **11.** 2.   **12.** 9.   **13.** 4.   **14.** 4.   **15.** 4.   **16.** 9.   **17.** 9.
18. 4.   **19.** 5.   **20.** 9.   **21.** 7.   **22.** 5.   **23.** 9.   **24.** 6.   **25.** 6_{12}.
26. 7_{12}.   **27.** 2_{12}.   **28.** 7_{12}.   **29.** Γ_{12}.   **30.** 9_{12}.   **31.** 7_{12}.   **32.** Σ_{12}.

EXERCISE 5

1. 24.   **2.** 25.   **3.** 34.   **4.** 35.   **5.** 49.   **6.** 77.   **7.** 38.   **8.** 236.
9. 595.   **10.** 77.   **11.** 368.   **12.** 713.   **13.** 526.   **14.** 108.
15. 4,075.   **16.** 2,797.   **17.** 6,973.   **18.** 708.   **19.** 4,092.
20. 5,386.   **21.** $709.   **22.** $900.   **23.** $12,898.   **24.** $720.
gain.   **25.** The one who bought the pawn ticket, for the pawn ticket was good for only $100 − 75 = $25.

EXERCISE 6

1. 83. **2.** 61. **3.** 313. **4.** 287. **5.** 470. **6.** 6,592. **7.** 5,927.
8. 3,500. **9.** 52. **10.** 63. **11.** 1,112. **12.** 956. **13.** 1,632.
14. 712. **15.** 1,608. **16.** 1,084.

EXERCISE 7

1. 536. **2.** 417. **3.** 860. **4.** 42,614. **5.** 491,886. **6.** 198.
7. 178. **8.** 4,019. **9.** 3,736. **10.** 3,256. **11.** 5,119. **12.** 17,158.
13. 48,434. **14.** 289,806. **15.** 279,483.
16. (**a**) 1,839 kw-hr; (**b**) Jul., 247 kw-hr; Aug., 254 kw-hr; Sept., 290 kw-hr; Oct., 326 kw-hr; Nov., 350 kw-hr; Dec., 372 kw-hr.
17. (**a**) Venezuela, 10 yrs., Ecuador, 11 yrs., Panama, 84 yrs.; (**b**) Venezuela, 352,150 sq. mi., Ecuador, 175,830 sq. mi., Panama 28,575 sq. mi.
18. $60 and the suit.
19. Since 9 birds eat 1 worm in 1 minute, only 9 birds will be needed.
20. It only took him 14 days to climb to the top, for on that day he climbed the last 11 feet.

EXERCISE 8

1. 12. **2.** 12. **3.** 35. **4.** 32. **5.** 15. **6.** 16. **7.** 24. **8.** 20.
9. 21. **10.** 6. **11.** 36. **12.** 45. **13.** 36. **14.** 49. **15.** 42.
16. 64. **17.** 72. **18.** 28. **19.** 81. **20.** 56. **21.** 48. **22.** 12.
23. 15. **24.** 36. **25.** 54. **26.** In 12 hours it strikes $1 + 2 + 3 + 4 + 5 + 6 + 7 + 8 + 9 + 10 + 11 + 12 = 78$ times. In 24 hours or 1 day, it strikes $2 \times 78 = 156$ times. **27.** 0.

EXERCISE 9

1. 68. **2.** 69. **3.** 36. **4.** 84. **5.** 208. **6.** 217. **7.** 126. **8.** 128.
9. 186. **10.** 155. **11.** 159. **12.** 246. **13.** 92. **14.** 600. **15.** 216.
16. 333. **17.** 336. **18.** 1,470. **19.** 1,537. **20.** 3,040. **21.** 5,544.
22. 5,762. **23.** 18,810. **24.** 43,672. **25.** 35,836. **26.** 217,761.
27. 306,804. **28.** 741,688. **29.** 873,144. **30.** 2,527,171.
31. 78,160,957. **32.** 876,814,809. **33.** $460. **34.** 142 miles.
35. $1,050. **36.** 11,446 gallons. **37.** $143.

EXERCISE 10

1. 550. **2.** 6,700. **3.** 769,000. **4.** 5,654. **5.** 1,848. **6.** 4,356.
7. 6,611. **8.** 3,996. **9.** 8,712. **10.** 9,585. **11.** 8,622. **12.** 3,154.
13. 15,164. **14.** 8,277. **15.** 19,459. **16.** 24,225. **17.** 30,463.

18. 32,305. **19.** 88,998. **20.** 45,854. **21.** 28,458. **22.** 27,089.
23. 23,364. **24.** 87,318. **25.** 2,410. **26.** 13,725. **27.** 9,250.
28. 37,000.

EXERCISE 11

1. 72. **2.** 45. **3.** 9,702. **4.** 8,448. **5.** 8,645. **6.** 994,008.
7. 993,006. **8.** 984,055. **9.** 8,667. **10.** 36,716. **11.** 10,116.
12. 118,594. **13.** 203,528. **14.** 105,567. **15.** 1,245,780.
16. 1,726,248. **17.** 102. **18.** 1,128. **19.** 2,646. **20.** 19,296.
21. 12,852. **22.** 15,995. **23.** 4,914. **24.** 8,841. **25.** 7,075.
26. 18,019. **27.** 24,932. **28.** 75,276. **29.** 173,862. **30.** 279,795.
31. 974,262. **32.** 425,445.

EXERCISE 12

1. 803. **2.** 19,437. **3.** 10,332. **4.** 9,101. **5.** 8,624. **6.** 993,012.
7. 34,563. **8.** 50,096. **9.** 151,866. **10.** 883,850. **11.** 625,646.
12. 2,795,235. **13.** $8.64. **14.** 2,856 peaches. **15.** 44,640 minutes.
16. 9,996 lines. **17.** 40,320 revolutions. **18.** 78,279 miles.
19. 882. **20.** The horsemen meet in 2 hours. The horsefly then flies 2
hours at 10 miles per hour or 20 miles.

EXERCISE 13

1. 2. **2.** 9. **3.** 2. **4.** 3. **5.** 4. **6.** 5. **7.** 6. **8.** 7. **9.** 8.
10. 9. **11.** 2. **12.** 3. **13.** 4. **14.** 5. **15.** 3. **16.** 4. **17.** 6.
18. 7. **19.** 8. **20.** 3. **21.** 4. **22.** 8. **23.** 3. **24.** 8. **25.** 5.
26. 7. **27.** 6. **28.** 6. **29.** 6. **30.** 5. **31.** 32. **32.** 31. **33.** 24.
34. 324. **35.** 132. **36.** 7, R = 3. **37.** 6, R = 7. **38.** 21, R = 2.
39. 31, R = 2. **40.** 40, R = 7. **41.** 71, R = 2.
42. 3,211, R = 1. **43.** 4,211, R = 3. **44.** 711, R = 2.
45. 3,225. **46.** 515. **47.** 4,263, R = 1. **48.** 138, R = 3.
49. 3,452, R = 5. **50.** 5,782, R = 2. **51.** 5,896, R = 2.
52. 3,472, R = 7. **53.** 14 days. **54.** 2 hours. **55.** $750.

EXERCISE 14

1. 2. **2.** 4. **3.** 6. **4.** 5. **5.** 3. **6.** 4. **7.** 5. **8.** 5. **9.** 4.
10. 6. **11.** 5. **12.** 7. **13.** 4. **14.** 7. **15.** 8. **16.** 9. **17.** 5.
18. 6. **19.** 9. **20.** 8. **21.** 56. **22.** 28. **23.** 35. **24.** 24.
25. 34. **26.** 78. **27.** 305. **28.** 407. **29.** 309. **30.** 303. **31.** 223.
32. 503, R = 69. **33.** 706, R = 4. **34.** 153, R = 73.
35. 108, R = 8. **36.** 314, R = 93. **37.** 206, R = 103.
38. 2,043, R = 374. **39.** 2,052, R = 154. **40.** 3,507, R = 475.

41. 1,136, R = 7,424.   **42.** 2,406, R = 167.

43. 25 × $50 = $1,250; 20 × $35 = $700; ($1, 250 + $350) − $700 = $900; 50 − 20 = 30; $900 ÷ 30 = $30.

44. (a) If 12 hats + 25 pairs of shoes = $235
 then   36 hats + 75 pairs of shoes = $705
 but   36 hats + 6 pairs of shoes = $222   hence
 _____
          69 pairs of shoes = $483   and
          1 pair of shoes = $483 ÷ 69 = $7.

 (b)   Then 36 hats + 6($7) = $222 so that
          36 hats = $222 − $42 = $180 and
          1 hat = $180 ÷ 36 = $5.

45. At the start of the second plane, the distance between the two planes is 180 miles. This distance is reduced by 240 − 180 = 60 miles per hour. Hence, it will take the second plane 180 ÷ 60 = 3 hours to overtake the first. Therefore, the time will be 1 P.M.

46. If 2 dogs + 5 puppies = 63 lbs., then 5 times the number of dogs and the number of puppies would weigh 5 times as much. Thus,

          10 dogs + 25 puppies = 135 lbs.
Similarly, if   5 dogs + 2 puppies = 126 lbs.
then   10 dogs + 4 puppies = 252 lbs.
but   10 dogs + 25 puppies = 315 lbs.
 _____
hence   21 puppies = 63 lbs. and
 (a)   1 puppy = 63 ÷ 21 = 3 lbs.
Then   5 dogs + 2(3) = 126
so that   5 dogs = 120 lbs. and
 (b)   1 dog = 120 ÷ 5 = 24 lbs.

47. Six dozen dozen = 6 × 12 × 12 = 864; half a dozen dozen = (12 × 12) ÷ 2 = 72.   **48.** 4 hours.

EXERCISE 15

1. 84. **2.** 47. **3.** 59. **4.** 723. **5.** 529. **6.** 637. **7.** 89. **8.** 76.
9. 57. **10.** 47, R = 35. **11.** 407, R = 15. **12.** 504, R = 52.
13. 423, R = 71. **14.** 607, R = 62. **15.** 503, R = 54.
16. 114 ÷ 3 = 38. **17.** (a) 504 ÷ 36 = 14 hours; (b) 504 ÷ 14 = 36 gallons. **18.** 118 × 25 = 2,950; 2,950 + 103 = 3,053; 3,053 − 50 = 3,003; 3,003 ÷ 7 = 429. **19.** 12. **20.** 15 days, because two of the ears are his own.

EXERCISE 16

1. 52. **2.** 31. **3.** 23. **4.** 53. **5.** 27. **6.** 62, R = 17.
7. 73, R = 15. **8.** 423, R = 9. **9.** 234, R = 47. **10.** 526, R = 26.

11. 21. **12.** 34. **13.** 64. **14.** 26. **15.** 57. **16.** 83, R = 52.
17. 47, R = 33. **18.** 503, R = 135. **19.** 120 × $4 = $480;
12 × $3 = $36; ($480 + $312) − $36 = $756; 120 − 12 = 108;
$756 ÷ 108 = $7.

20. (a) If $\quad$ 5 steers + $\quad$ 4 horses cost $\quad$ $410
$\qquad$ then 15 steers + 12 horses cost $1,230
$\qquad$ but $\quad$ <u>15 steers + $\quad$ 6 horses cost $\quad$ $990 </u> hence

$\qquad\qquad\qquad$ 6 horses cost $\quad$ $240 and
$\qquad\qquad\qquad$ 1 horse costs $\quad$ $240 ÷ 6 = $40.

$\quad$ **(b)** Then 5 steers + 4 ($40) costs $410 so that
$\qquad\qquad$ 5 steers cost $410 − $160 = $250 and
$\qquad\qquad$ 1 steer costs $250 ÷ 5 = $50.

21. Every hour the distance between the trains is reduced by 57 + 32 = 89 miles, hence it will take them 178 ÷ 89 = 2 hours.

22. 406 barrels.

23. One-half of $\overset{\times}{\underset{\mathsf{l\,l}}{}}$ is $\mathsf{V}\,\mathsf{l}\,\mathsf{l}$.

24. One-half of $\mathsf{l}\overset{\mathsf{l}}{\times}$ is $\mathsf{l}\mathsf{V}$.

25. Give one of the boys the box with an apple in it.

EXERCISE 17

1. 1, 2, 3, 5, 7, 11, 13, 17, 19, 23, 29, 31, 37, 41, 43, 47, 53, 59, 61, 67, 71, 73, 79, 83, 89, 97. **2.** 2 × 3 × 7. **3.** 3 × 5 × 5. **4.** 3 × 29.
5. 2 × 47. **6.** 2 × 2 × 2 × 2 × 2 × 2 × 3. **7.** 5 × 37.
8. 13 × 31. **9.** 3 × 67. **10.** 2 × 5 × 7 × 11. **11.** 3 × 3 × 3 × 5 × 13. **12.** 2 × 7 × 11 × 13. **13.** 2 × 2 × 3 × 3 × 5 × 7.
14. 2 × 2 × 2 × 3 × 3 × 3 × 7. **15.** 2 × 5 × 3 × 3 × 11 × 17.

EXERCISE 18

1. 9. **2.** 18. **3.** 8. **4.** 75. **5.** 54. **6.** 6. **7.** 18. **8.** 8. **9.** 16.
10. 45. **11.** 8. **12.** 20. **13.** 8. **14.** 25. **15.** 36. **16.** 18.
17. 12 yards. **18.** 6 yards. **19.** $5. **20.** Two; she took 1 apple and left 1 apple.

EXERCISE 19

1. 1,620. **2.** 72. **3.** 910. **4.** 5,616. **5.** 43,875. **6.** 1,260.
7. 4,930. **8.** 35,244. **9.** 42,000. **10.** 29,816,640. **11.** 4,680.
12. 24,200. **13. (a)** 42 days; **(b)** 1st, 6 trips; 2nd, 3 trips; 3rd, 2 trips. **14.** 180 feet. **15.** The smallest number is the L. C. M.
(2, 3, 4, 5, 6, 7, 8, 9, 10) less 1; but, L. C. M. (2, 3, 4, 5, 6, 7, 8, 9, & 10) = 2,520; therefore the number is 2,520 − 1 = 2,519.

EXERCISE 20

1. 18. **2.** 36. **3.** 128. **4.** 3,375. **5.** 1,764. **6.** 1,764. **7.** 252.
8. 2,000. **9.** $4^5 = 1,024$. **10.** $5^5 = 3,125$. **11.** $35 \times 17 \times 10^9 =$
595,000,000,000. **12.** $41 \times 23 \times 10^9 = 943,000,000,000$. **13.** $81 \times$
$9 \times 10^8 = 72,900,000,000$. **14.** $76 \times 18 \times 10^7 = 13,680,000,000$.
15. $265 \times 41 \times 10^7 = 108,650,000,000$. **16.** 75. **17.** 27. **18.** 19.
19. 63. **20.** 84. **21.** 104. **22.** 207. **23.** 349. **24.** 381. **25.** 758.
26. $1 \div \sqrt{1} = 1$.

EXERCISE 21

1. (a) 8; (b) 32; (c) 48. **2.** $3\frac{3}{3}$; $5\frac{5}{5}$; and so on. **3.** (a) 2;
(b) 4; (c) 5. **4.** Four sevenths, three sevenths. **5.** (a) $\frac{15}{9}$, $\frac{79}{78}$;
(b) $\frac{8}{15}$, $\frac{13}{28}$, $\frac{3}{5}$; (c) $\frac{23}{23}$, $\frac{12}{12}$; (d) $\frac{15}{9}$; (e) $\frac{13}{28}$. **6.** Each boy eats $2\frac{2}{3}$
bars so that the third boy gets $\frac{1}{3}$ of a bar from the boy with 3 bars and
$\frac{7}{3}$ or $2\frac{1}{3}$ bars from the boy with 5 bars; hence, he pays 7 cents to the boy
with 5 bars and 1 cent to the boy with 3 bars. **7.** (a) $\frac{2}{7}$; (b) $\frac{4}{11}$;
(c) $\frac{12}{25}$; (d) $\frac{5}{19}$. **8.** (a) $\frac{2}{5}$; (b) $\frac{3}{8}$; (c) $\frac{12}{13}$; (d) $\frac{53}{8} = 6\frac{5}{8}$.
9. (a) $\frac{2}{7}$; (b) $\frac{2}{5}$; (c) $\frac{2}{4} = \frac{1}{2}$; (d) $\frac{2}{8} = \frac{1}{4}$. **10.** (a) $\frac{3}{13}$; (b) $\frac{3}{9} = \frac{1}{3}$;
(c) $\frac{3}{18} = \frac{1}{6}$; (d) $\frac{3}{7}$;

EXERCISE 22

1. (a) Fraction is multiplied by 2; (b) fraction is divided by 3.
2. (a) Fraction is multiplied by 4; (b) fraction is divided by 4.
3. (a) Fraction is divided by 2; (b) fraction is multiplied by 3.
4. (a) Fraction is multiplied by 3; (b) fraction is divided by 3.
5. (a) None in value; (b) none in value. **6.** 6. **7.** 3. **8.** 9.

9. 1. **10.** 5. **11.** 3. **12.** 14. **13.** 8. **14.** $\frac{3}{14} \times 7 = \frac{3 \times 7}{14} =$

$\frac{3}{14 \div 7} = \frac{3}{2}$; $\frac{2}{21} \times 7 = \frac{2 \times 7}{21} = \frac{2}{21 \div 7} = \frac{2}{3}$; $\frac{5}{35} \times 7 =$

$\frac{5 \times 7}{35} = \frac{5}{35 \div 7} = \frac{5}{5} = 1$; **15.** $\frac{6}{7} \div 3 = \frac{6 \div 3}{7} = \frac{6}{7 \times 3} = \frac{2}{7}$;

$\frac{12}{17} \div 3 = \frac{12 \div 3}{17} = \frac{12}{17 \times 3} = \frac{4}{17}$; $\frac{15}{28} \div 3 = \frac{15 \div 3}{28} = \frac{15}{28 \times 3} =$

$\frac{5}{28}$.

EXERCISE 23

1. $\frac{3}{2}$. **2.** $\frac{6}{5}$. **3.** $\frac{9}{4}$. **4.** $\frac{31}{5}$. **5.** $\frac{14}{3}$. **6.** $\frac{35}{4}$. **7.** $\frac{41}{6}$. **8.** $\frac{27}{8}$. **9.** $\frac{33}{7}$.
10. $\frac{38}{3}$. **11.** $\frac{63}{4}$. **12.** $\frac{80}{7}$. **13.** $3\frac{1}{2}$. **14.** $4\frac{2}{3}$. **15.** $6\frac{1}{4}$. **16.** $9\frac{3}{5}$.

17. $9\frac{4}{9}$. **18.** $12\frac{2}{5}$. **19.** $10\frac{2}{11}$. **20.** $6\frac{5}{14}$. **21.** $3\frac{4}{13}$. **22.** $8\frac{7}{12}$.
23. $53\frac{9}{17}$. **24.** $42\frac{4}{25}$. **25.** $2 + \frac{22}{22} - \frac{2}{2} = 2$.

EXERCISE 24

1. $\frac{4}{20}$. **2.** $\frac{8}{12}$. **3.** $\frac{6}{39}$. **4.** $\frac{48}{102}$. **5.** $\frac{54}{63}$. **6.** $\frac{48}{174}$. **7.** $\frac{184}{200}$. **8.** $\frac{355}{415}$.
9. $\frac{3}{9}$. **10.** $\frac{2}{9}$. **11.** $\frac{7}{17}$. **12.** $\frac{4}{3}$. **13.** $\frac{3}{18}$. **14.** $\frac{13}{52}$. **15.** $\frac{17}{52}$. **16.** $\frac{11}{85}$.
17. $\frac{2}{5}$. **18.** $\frac{3}{4}$. **19.** $\frac{2}{7}$. **20.** $\frac{3}{11}$. **21.** $\frac{2}{3}$. **22.** $\frac{5}{7}$. **23.** $\frac{5}{9}$. **24.** $\frac{7}{9}$.
25. $\frac{3}{5}$. **26.** $\frac{2}{3}$. **27.** $\frac{7}{11}$. **28.** $\frac{11}{13}$. **29.** $\frac{1}{8}$. **30.** $\frac{3}{56}$. **31.** $\frac{1}{13}$. **32.** $\frac{2}{5}$.
33. 3. **34.** $\frac{10}{3}$.

EXERCISE 25

1. 21. **2.** 35. **3.** 30. **4.** 24. **5.** 18. **6.** 24. **7.** 81. **8.** 36.
9. 72. **10.** 270. **11.** 300. **12.** 180. **13.** 1,680. **14.** 120. **15.** 72.
16. 360. **17.** 70,560. **18.** 14,560.

EXERCISE 26

1. $\frac{5}{6}$. **2.** $\frac{2}{5}$. **3.** $\frac{8}{9}$. **4.** $\frac{7}{6} = 1\frac{1}{6}$. **5.** $\frac{1}{20}$. **6.** $\frac{19}{78}$. **7.** $\frac{17}{30}$. **8.** $15\frac{6}{11}$.
9. $8\frac{11}{14}$. **10.** $3\frac{15}{28}$. **11.** $10\frac{7}{10}$ **12.** $3\frac{3}{4}$. **13.** $8\frac{101}{220}$. **14.** $7\frac{2}{3}$.
15. $2\frac{11}{36}$. **16.** $\frac{13}{120}$. **17.** $1\frac{15}{28}$. **18.** $\frac{19}{12}$. **19.** $\frac{7}{24}$. **20.** $\frac{17}{24}$. **21.** $\frac{199}{225}$.
22. $2\frac{329}{360}$. **23.** $4\frac{14}{15}$. **24.** $6\frac{11}{84}$. **25.** $3\frac{2}{3}$. **26.** $\frac{667}{1,680}$. **27.** $\frac{3,173}{3,780}$.
28. $\frac{69}{220}$. **29.** $1\frac{3}{4}$. **30.** $13\frac{13}{42}$. **31.** $7\frac{73}{304}$. **32.** $\frac{35}{70} + \frac{148}{296} = 1$.
33. One-half of its weight is 10 lbs; hence, its weight $= 2 \times 10 = 20$ lbs. **34.** (a) $200\frac{83}{120}$ lbs.;   (b) $150\frac{49}{72}$ lbs.;   (c) $50\frac{1}{90}$ lbs.
35. $206\frac{23}{120}$ bushels.

EXERCISE 27

1. $\frac{1}{3}$. **2.** $\frac{5}{14}$. **3.** $\frac{10}{19}$. **4.** $\frac{1}{4}$. **5.** $\frac{2}{13}$. **6.** $\frac{2}{3}$. **7.** $\frac{1}{3}$. **8.** $\frac{1}{12}$. **9.** 7.
10. $\frac{1}{7}$. **11.** $\frac{10}{17}$. **12.** 1. **13.** $\frac{1}{3}$. **14.** $2\frac{1}{3}$. **15.** $\frac{1}{3}$. **16.** $\frac{11}{16}$.
17. 1 minute. **18.** $87\frac{2}{5}$ gallons. **19.** \$27. **20.** \$628.

EXERCISE 28

1. $\frac{1}{15}$. **2.** $\frac{3}{7}$. **3.** $2\frac{1}{2}$. **4.** $4\frac{3}{16}$. **5.** $\frac{3}{5}$. **6.** $\frac{1}{4}$. **7.** $\frac{16}{69}$. **8.** $2\frac{5}{8}$.
9. $1\frac{5}{27}$. **10.** $\frac{2}{3}$. **11.** $\frac{1}{2}$. **12.** $\frac{2}{19}$. **13.** $\frac{2}{7}$. **14.** $\frac{1,859}{8,880}$. **15.** 5.
16. $\frac{1}{21}$. **17.** $\frac{9}{10}$ of a pound $= \frac{1}{10}$ of its weight; therefore, weight $= 10 \times \frac{9}{10} = 9$ lbs. **18.** $3\frac{1}{13}$ hours. **19.** 30 hours.
20. First number $= \frac{1}{4}$(number $+ 1$) hence, 4 times first number $=$ number $+ 1$; then 3 times first number $= 1$, and first number $= \frac{1}{3}$. Now second number $= \frac{1}{4}$(second number) $\times$ (second number $+ 1$) hence, $4 \times$ second number $=$ second number $\times$ (second number $+ 1$). So that $4 =$ second number $+ 1$, and $3 =$ second number. Therefore, $\frac{1}{3} \times 3 = 1$.

21. First year, he had $\frac{3}{3}$ of original amount; second year, he had $\frac{4}{3}$ of original amount; third year, he had $\frac{4}{3}$ of $\frac{4}{3} = \frac{4}{3} \times \frac{4}{3} = \frac{16}{9}$ of original amount; and fourth year, he had $\frac{4}{3}$ of $\frac{16}{9} = \frac{4}{3} \times \frac{16}{9} = \frac{64}{27}$ of original amount. If $\frac{64}{27}$ of original amount = \$64,000, then original amount = \$64,000 $\times \frac{27}{64}$ = \$27,000.

22. $131\frac{1}{4}$ lbs. of 75¢ candy and $218\frac{3}{4}$ lbs. of 35¢ candy.

23. $\frac{600}{10,000} = \frac{3}{50}$; $1 + \frac{3}{50} = \frac{53}{50}$; $1 + \frac{1}{8} = \frac{9}{8}$; $\frac{9}{8} - \frac{53}{50} = \frac{13}{200}$; $1 - \frac{1}{5} = \frac{4}{5}$; $\frac{53}{50} - \frac{4}{5} = \frac{13}{50}$ or $\frac{52}{200}$; $13 + 52 = 65$. **(a)** $\frac{52}{65}$ of \$10,000 = \$8,000; **(b)** $\frac{13}{65}$ of \$10,000 = \$2,000.

24. Travels 1 mile in $\frac{1}{20}$ of an hour going. Travels 1 mile in $\frac{1}{30}$ of an hour returning. Both ways, travels 1 mile in $\frac{1}{2}$ $(\frac{1}{20} + \frac{1}{30}) = \frac{1}{24}$ of an hour. Therefore, rate is 24 miles per hour.

25. Time spent for lunch = distance traveled by minute hand + distance traveled by hour hand = one complete revolution. Since minute hand travels 12 times faster than hour hand, then distance traveled by minute hand is $\frac{12}{13}$, and distance traveled by hour hand is $\frac{1}{13}$ of a revolution. But $\frac{12}{13} \times 60 = 55\frac{5}{13}$ minutes. Thus time spent for lunch is $55\frac{5}{13}$ minutes. Since minute hand travels 12 times faster than hour hand, then distance between them was $\frac{11}{12}$ of distance of minute hand from XII or zero point. **(a)** Hence $\frac{11}{12}$ of distance = $\frac{1}{13}$ of a revolution, or distance = $\frac{1}{13} \times \frac{12}{11} = \frac{12}{143}$ of 60 = $5\frac{5}{143}$ minutes past twelve; **(b)** $5\frac{5}{143} + 55\frac{5}{13} = 60\frac{60}{143}$ or $\frac{60}{143}$ past one o'clock.

EXERCISE 29

1. 7. **2.** 0.32. **3.** 20.34. **4.** 2. **5.** 0.07. **6.** 2.301. **7.** 0.3215. **8.** 340. **9.** 0.157. **10.** 0.072305. **11.** 2,682.01. **12.** 0.105312. **13.** 10. **14.** 10. **15.** 100. **16.** 100. **17.** 100. **18.** 10,000. **19.** 100,000. **20.** 1,000,000.

EXERCISE 30

1. 0.91. **2.** 1.5932. **3.** 2.970456. **4.** 0.281. **5.** 4.8422. **6.** 2.157. **7.** 0.55683. **8.** 2.2173. **9.** 2.36623. **10.** 4.15525. **11.** 5.11277. **12.** 3.554887. **13.** 5.83379812. **14.** 4.290285. **15.** 9.643426. **16.** 3.508325. **17.** \$58.66. **18.** \$168.43. **19.** 0.47317 gm. **20.** \$601.008716 or \$601.01.

EXERCISE 31

1. 0.28. **2.** 0.119. **3.** 233.9358. **4.** 2.710268. **5.** 0.03165698. **6.** 0.211459584. **7.** 0.95625945. **8.** 0.5073. **9.** 1,138.301901. **10.** 43.45309956. **11.** 3.375 inches. **12.** 30,798.0288 lbs. **13.** **(a)** 138.567 gallons; **(b)** \$51.6162075 or \$51.62. **14.** \$495.7677919494 or \$495.77. **15.** \$650.174464 or \$650.17.

EXERCISE 32

1. 0.016. **2.** 0.030. **3.** 24.380. **4.** 1.803. **5.** 2.192. **6.** 6.627.
7. 0.350. **8.** 0.538. **9.** 0.236. **10.** 4.632. **11.** 10.279.
12. 12.438. **13.** 2,342.857. **14.** 5.267. **15.** (a) 8; (b) 30.608 lbs.
16. 1,400 miles. **17.** 8.79 minutes. **18.** 0.25. **19.** 29.02 minutes.
20. $20.59.

EXERCISE 33

1. $\frac{1}{5}$. **2.** $\frac{1}{25}$. **3.** $\frac{1}{4}$. **4.** $\frac{9}{20}$. **5.** $\frac{5}{16}$. **6.** $\frac{3}{8}$. **7.** $\frac{5}{12}$. **8.** $\frac{963}{40}$. **9.** $\frac{1,201}{400}$.
10. $\frac{41}{40}$. **11.** $\frac{83}{6}$. **12.** $\frac{68}{7}$. **13.** 0.2. **14.** 0.625. **15.** 0.6. **16.** 0.15.
17. 0.025. **18.** 0.95. **19.** 0.833. **20.** 0.416. **21.** 14.286.
22. 0.636. **23.** 0.385. **24.** 0.471. **25.** 0.208. **26.** 0.167.
27. 0.467.

EXERCISE 34

1. 6.2137×10^{-1}. **2.** 1.8633×10^5. **3.** 4.774×10^{-10}.
4. 6.06×10^{23}. **5.** 4.1834×10^7. **6.** 12. **7.** 22. **8.** 0.856.
9. 3.6. **10.** 48.5. **11.** 404. **12.** 10. **13.** 16. **14.** 10. **15.** 32.8.
16. 1,130. **17.** 4.30. **18.** 27.0. **19.** 4.175. **20.** 155.

EXERCISE 35

1. 1.9835. **2.** 7.686. **3.** 11.4674. **4.** 8.14. **5.** 3.39. **6.** 2.183.
7. 545. **8.** 3,675.04. **9.** 1,527.526. **10.** 10,285.8.

EXERCISE 36

1. 15.81. **2.** 8.31. **3.** 19.47. **4.** 34.93. **5.** 89.11. **6.** 3.67.
7. 29.59. **8.** 41.84. **9.** 0.86. **10.** 40.04. **11.** 167.81. **12.** 104.53.
13. $1.18. **14.** 84.82 yds. **15.** $1,207.75947066 or $1,207.76.
16. $104.97. **17.** $28,226.95. **18.** Since the integer is a perfect
square, the right-hand group must be 76, and the unit figure of the root
must be 4 for 4 × 4 = 16, giving 6, which is the last figure of the dividend. Hence the last trial divisor is 1376 ÷ 4 = 344. It follows that
4 must have been annexed to 34, which must be twice the root already
found. Since 34 ÷ 2 = 17, the integer is 174 and its square is 174 ×
174 = 30,276.
19. EL is one-third of TW(EL)VE and EVEN is four-fifths of S(EVEN).
EL + EVEN = ELEVEN.
20. Fill the three gallon measure and pour it into the 5 gallon measure.
Fill the 3 gallon measure again and pour it into the 5 gallon measure
until the 5 gallon measure is full. There is now 1 gallon left in the 3

gallon measure. Empty the 5 gallon measure and pour the 1 gallon into it. Fill the 3 gallon measure again and pour it into the 5 gallon measure obtaining exactly 4 gallons.

EXERCISE 37

1. 5. **2.** $\frac{7}{8}$. **3.** $\frac{3}{4}$. **4.** $\frac{7}{3}$. **5.** $\frac{3}{4}$. **6.** $\frac{7}{4}$. **7.** $\frac{4}{125}$. **8.** $\frac{1}{20}$. **9.** $\frac{7}{3,200}$.
10. $\frac{23}{5}$.

EXERCISE 38

1. 10. **2.** 6. **3.** 32. **4.** 3. **5.** 6. **6.** $7\frac{1}{8}$. **7.** 15. **8.** 3.
9. 16. **10.** 3,150 cubic feet. **11.** $231. **12.** $56\frac{1}{4}$ feet. **13.** 250 lbs.
of cement, 750 lbs. of sand, 1,500 lbs. of gravel. **14.** $30,320,
$24,003.33, $11,370, $10,106.67. **15.** 184 miles.

EXERCISE 39

1. (b) $10:2 = 20:4, 10:8 = 20:16$. **2. (b)** $3:1 = 15:5, 3:2 = 15:10$.
3. (b) $9:6 = 3:2, 9:3 = 3:1$. **4. (b)** $40:15 = 16:6, 40:25 = 16:10$.
5. (b) $28:25 = 56:50, 28:3 = 56:6$. **6. (b)** $24:8 = 15:5, 24:16 = 15:10$. **7. (b)** $235:160 = 47:32, 235:75 = 47:15$. **8. (b)** $108:57 = 180:95, 108:51 = 180:85$. **9. (b)** $90:51 = 150:85, 90:39 = 150:65$.
10. 8. **11.** 18. **12.** 16. **13.** 15. **14.** 14. **15.** 21. **16.** 20.
17. 22. **18.** 34. **19.** $34\frac{1}{2}$ hours. **20.** 46 lbs.

EXERCISE 40

1. 1,800 revolutions. **2.** 8 days. **3.** 440 yards. **4.** 400 miles per
hour. **5.** 18 feet per second. **6.** $29,866.67. **7.** $7,623. **8.** $3,696.
9. 8 days. **10.** 40 men.

EXERCISE 41

1. (a) 0.0625; **(b)** $\frac{1}{16}$. **2. (a)** 0.08; **(b)** $\frac{2}{25}$. **3. (a)** 0.4; **(b)** $\frac{2}{5}$.
4. (a) 0.125; **(b)** $\frac{1}{8}$. **5. (a)** 0.375; **(b)** $\frac{3}{8}$. **6. (a)** 0.625; **(b)** $\frac{5}{8}$.
7. (a) $0.1428\frac{4}{7}$; **(b)** $\frac{1}{7}$. **8. (a)** $0.0041\frac{2}{3}$; **(b)** $\frac{1}{240}$. **9. (a)** $0.004\frac{4}{9}$;
(b) $\frac{1}{225}$. **10. (a)** $0.33\frac{1}{3}$; **(b)** $\frac{1}{3}$. **11. (a)** 0.0175; **(b)** $\frac{7}{400}$.
12. (a) 0.068; **(b)** $\frac{17}{250}$. **13. (a)** 0.0375; **(b)** $\frac{3}{80}$. **14. (a)** 0.16375;
(b) $\frac{131}{800}$. **15. (a)** 0.171875; **(b)** $\frac{11}{64}$. **16.** 45%. **17.** $37\frac{1}{2}$%.
18. $12\frac{1}{2}$%. **19.** $6\frac{1}{4}$%. **20.** $\frac{7}{8}$%. **21.** $101\frac{2}{5}$%. **22.** $162\frac{1}{2}$%.
23. 55%. **24.** 60%. **25.** $87\frac{1}{2}$%. **26.** 5%. **27.** $21\frac{7}{8}$%. **28.** $31\frac{1}{4}$%.
29. $21\frac{1}{2}$%. **30.** $19\frac{4}{5}$%.

EXERCISE 42

1. 21.6. **2.** 10.8. **3.** 18.75. **4.** $18.75. **5.** $43.75. **6.** 4.21875.
7. 7.5. **8.** 3.25. **9.** 20%. **10.** 30%. **11.** 4%. **12.** 7.2%.
13. $12\frac{1}{2}\%$. **14.** 23.4%. **15.** 150. **16.** 2,200. **17.** $2,691.67.
18. 3,800. **19.** 84,000. **20.** 3,560. **21.** $548.57. **22.** $15.
23. (a) Rent—$1,900, Food—$2,375, Clothing—$950, Medical—$475,
Insurance—$190, Amusements—$855, Taxes and miscellaneous—
$1,805; **(b)** $950. **24. (a)** 23.8%; **(b)** 57.2%; **(c)** 19%.
25. 82.9%. **26.** 0.45%. **27.** $140,960. **28.** $299.50. **29.** $4.60.
30. $95,000.

EXERCISE 43

1. $1,303.05. **2. (a)** 1,814; **(b)** $0.49. **3. (a)** $570; **(b)** 40%.
4. (a) $5,521.21; **(b)** 19%. **5.** $583.53. **6.** $18,056.10.
7. $136.85. **8.** $2,915. **9.** 1,268. **10.** 424. **11. (a)** 1,134.
(b) $0.50. **12. (a)** $32.83. **(b)** 28%. **13. (a)** $1,093.23.
(b) $25\frac{3}{16}\%$. **14. (a)** $1,211.18. **(b)** 37.88%. **15. (a)** $4,708.87.
(b) 18.27%. **16.** 7%. **17.** 10%. **18.** 3%. **19. (a)** $324.68.
(b) $356.79. **20. (a)** $113.71. **(b)** $135.36.

Index